15. Band, 3. Heft

Fortschritte der chemischen Forschung
Topics in Current Chemistry

Herausgeber:

Prof. Dr. A. Davison   Department of Chemistry, Massachusetts Institute
                          of Technology, Cambridge, MA 02139, USA

Prof. Dr. M. J. S. Dewar   Department of Chemistry, The University of Texas
                          Austin, TX 78712, USA

Prof. Dr. K. Hafner   Institut für Organische Chemie der TH
                          6100 Darmstadt, Schloßgartenstraße 2

Prof. Dr. E. Heilbronner   Physikalisch-Chemisches Institut der Universität
                          CH-4000 Basel, Klingelbergstraße 80

Prof. Dr. U. Hofmann   Institut für Anorganische Chemie der Universität
                          6900 Heidelberg 1, Tiergartenstraße

Prof. Dr. K. Niedenzu   University of Kentucky, College of Arts and Sciences
                          Department of Chemistry, Lexington, KY 40506, USA

Prof. Dr. Kl. Schäfer   Institut für Physikalische Chemie der Universität
                          6900 Heidelberg 1, Tiergartenstraße

Prof. Dr. G. Wittig   Institut für Organische Chemie der Universität
                          6900 Heidelberg 1, Tiergartenstraße

Schriftleitung:

Dipl.-Chem. F. Boschke   Springer-Verlag, 6900 Heidelberg 1, Postfach 1780

Springer-Verlag   6900 Heidelberg 1 · Postfach 1780
                          Telefon (0 62 21) 4 91 01 · Telex 04-61723
                          1000 Berlin 33 · Heidelberger Platz 3
                          Telefon (03 11) 82 20 01 · Telex 01-83319

Springer-Verlag   New York, NY 10010 · 175, Fifth Avenue
New York Inc.   Telefon 673-26 60

Allene-Olefin and Allene-Allene Cycloadditions
Methylenecyclobutane and 1,2-Dimethylenecyclobutane Degenerate Rearrangements

Prof. J. E. Baldwin and Mr. R. H. Fleming

Department of Chemistry, University of Oregon, Eugene, OR 97403, USA

Contents

I. Introduction

Allene-olefin and allene-allene thermal cycloadditions provide useful synthetic routes to methylenecyclobutanes and 1,1-dimethylenecyclobutanes. The historical development, many practical exploitations, and mechanistic assessments of these reactions were summarized by Roberts and Sharts in 1962 [98] and covered in part in subsequent reviews related to allene chemistry [23,54,79,81,93,125,141].

$$CH_2 = C = CH_2 \quad + \quad CH_2 = CH_2 \quad \longrightarrow$$

$$CH_2 = C = CH_2 \quad + \quad CH_2 = C = CH_2 \quad \longrightarrow$$

Within the past few years *degenerate valence isomerizations* of methylenecyclobutane and 1,2-dimethylenecyclobutane have been detected.

$$H_2 \quad CD_2 \quad \rightleftarrows \quad D_2 \quad CH_2$$

$$H_2 \quad CD_2 \quad \rightleftarrows \quad D_2 \quad CH_2$$

New kinetic and stereochemical information on these two cyclo-additions and two degenerate molecular rearrangements has become available, and a comprehensive theory relating some general postulates on conservation of orbital symmetry during concerted chemical processes to their stereochemical features has been advanced and impressively developed [113,142].

These new developments place allene-olefin and allene-allene thermal cycloadditions in a new context framed by the questions: Are there *reactive intermediates* separating reactants and products in the cyclo-additions, or in the degenerate rearrangements? Might there be a common reactive intermediate for the allene-olefin addition and the methyl-enecyclobutane rearrangement, and another for the allene-allene addition and the 1,2-dimethylenecyclobutane rearrangement? How may the experi-mentally observed stereochemical features of these four processes be reconciled with orbital symmetry theory?

In focusing on these questions, we neglect several closely related topics that would merit inclusion in a review article of broader scope: resolutions and determinations of absolute configuration for allenes [27,29,34,69,118,133,134]; oligomers of allene [68,88,90,91,123,138–140]; other cycloadditions involving allenes [1,12,14,17,21,22,46,60,84,96,97,117,131]; stereoselectivity in oxymercurations and closely related additions to allenes [2,3,70,115,116,135–137]; thermal cycloreactions of allenes and allene dimers [55,74,120,127,129]; and other formally analogous cycloaddition, degenerate rearrangement couples such as allene plus carbene and the degenerate methylenecyclopropane rearrangement [32,42,48,50,92].

II. Allene-Olefin Cycloadditions

Kiefer and Okamura [73] demonstrated stereoselectivity with respect to the olefinic component in an allene-olefin cycloaddition. 1,1-Dimethyl-allene with dimethyl fumarate at 160—180 °C gave two cycloadducts, both retaining the *trans* disposition of ester functions with better than 99% stereoselectivity; dimethyl maleate gave the isomeric *cis* adducts with

high, but not precisely determinable selectivity. The *trans* olefin reacted
faster by a factor of 9.3 at 170 °C than the maleate.

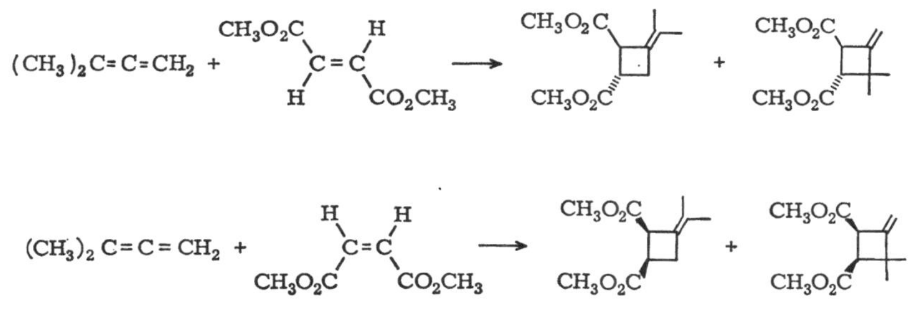

Kiefer and Khaleque determined that the rate of cycloaddition of 1,1-dimethylallene with acrylonitrile varied by less than a factor of two in solvents ranging in polarity between cyclohexane and dimethyl sulfoxide [73].

Dolbier and Dai [43] found through intramolecular competition experiments with 1,1-dideuterioallene and acrylonitrile at 206—225 °C that the carbon bearing deuterium atoms was found predominantly at the exocyclic position. The observed product ratio, interpreted in terms of a partitioning of an intermediate, gave k_H/k_D 1.13—1.21.

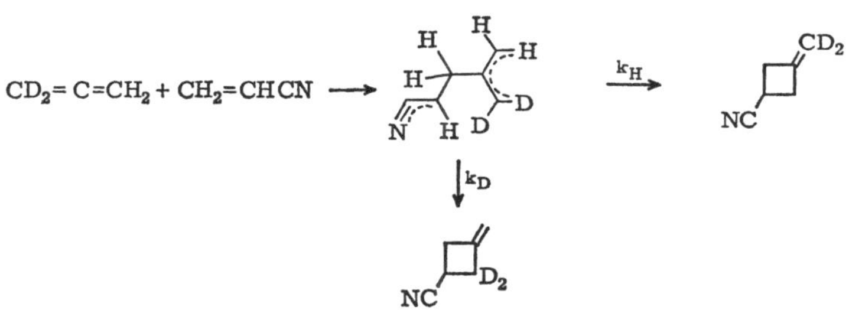

We have found through intramolecular competition experiments, reacting mixtures of allene and tetradeuterioallene with acrylonitrile at 200 °C to low conversion, a slight kinetic preference for the unlabeled allene: $k_H/k_D = 1.06$ [7]. No conscionable manipulation of statistical factors or postulation of probable errors can make this ratio equivalent to the 1.13—1.21 observed by Dolbier and Dai, and therefore the overall reaction must proceed by way of an intermediate. Kinetic isotope effects starting from allene and reaching a rate determining transition state are experimentally distinct from the isotope effects starting from an intermediate and reaching product determining transition states.

Various mixtures of acrylonitrile and α-deuterioacrylonitrile with allene in benzene at 200 °C gave 3-cyanomethylenecyclobutane with deuterium content appropriate for $k_H/k_D = 1.00 \pm 0.02$ [7].

The stereoselectivity of the cycloaddition with respect to the allene component was learned through study of the reaction of R-(-)-1,3-dimethylallene [69,133,136] and acrylonitrile [8]. The reaction gave *four cycloadducts* in approximately equal proportions, all having a predominance of the R configuration at C(2) of the ethylidenecyclobutanes.

This stereochemical result was interpreted through a model postulating least hindered approach of olefin toward allene and rotation of the methylene group to form the allylic moiety of the intermediate only in the sense specified by orbital symmetry requirements. The p orbital of the allene molecule originally paired with the p orbital on C(2) participating in the initial bonding contact with the olefin rotates towards the substituted carbon of the olefin in accord with orbital symmetry control if the reaction were a $\pi^2 s + \pi^2 s + \pi^2 s$ process [142]. The rotation is so controlled, but the reaction doesn't continue to follow the $\pi^2 s + \pi^2 s + \pi^2 s$ scenario; an intermediate is formed able to bond C(α) with either C(1) or C(3) in a disrotatory fashion.

The four ethylidenecyclobutanes derived through this formalism are precisely the four observed.

The minimum interpretation of the experimental results [8] rules out nondissymmetric intermediates or intermediates free to achieve nondissymmetric status on a time-averaged basis. The model proposed is not established, of course, but it may have utility for planning further work. It accommodates all available stereochemical and kinetic isotopic data, it would predict optically inactive adducts from optically active 1,3-dideuterioallene and acrylonitrile [41], because steric hindrance would probably not discriminate between alternative modes of approach, and a higher degree of stereoselectivity from active 1,3-di-*tert*-butylallene [27] than from 1,3-dimethylallene in such cycloadditions.

The product ratios observed for additions between allene and chlorotrifluoroethylene or 1,1-dichloro-2,2-difluoroethylene [126], 85:15 in favor of the 3-chloro isomer and 95:5 for the 3,3-dichloro adduct, can't be interpreted unambiguously. If both adducts in a set stem from a common type of intermediate, then the product ratios indicate how different that intermediate is from the sorts of diradicals implicated in cycloadditions between, say, 1,1-dichloro-2,2-difluoroethylene and the 2,4-hexadienes [16]. But the two adducts may arise through completely independent, dissimilar reaction mechanisms.

III. Methylenecyclobutane Rearrangements

Kinetic investigations of the spiropentane isomerization to methylenecyclobutane and decomposition to allene and ethylene by Burkhardt and Swinehart [30] and of thermal decomposition of methylenecyclobutane by Chesick [31] led to the recognition that a common intermediate might be involved. Other workers provided additional kinetic data [28,47].

Chesick [31] proposed the intermediate might make possible a degenerate type of rearrangement by which methylenecyclobutane would be converted into itself. This type of rearrangement was soon detected in the ethylidenecyclopropane interconversion with 2-methylmethylenecyclopropane

[32]), and in methylenecyclobutane itself using deuterium labeling by Doering and Gilbert [41].

With the hypothesis that allene-olefin cycloadditions and methylenecyclobutane rearrangements have a common intermediate comes the prediction that common stereochemical modes for the various rotations involved will prevail. The cycloaddition model outlined above involves disrotatory ring closure with the same sense of rotation about C(4)C(5) and C(5)C(2). For the methylenecyclobutane rearrangement, the same stereochemistry is anticipated for the reverse reaction.

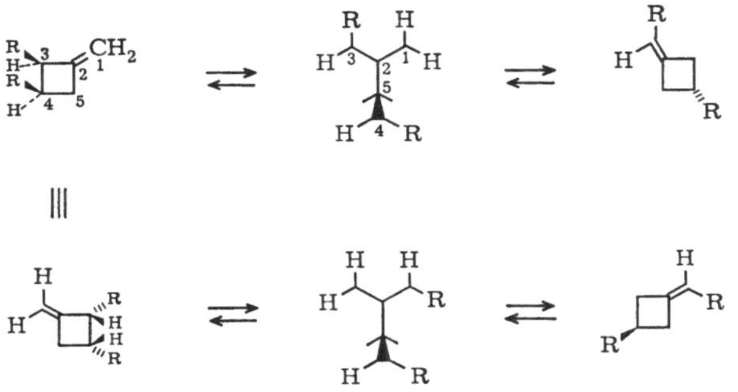

The two rearrangement products from the given substituted methylenecyclobutane are seen to be related as mirror images, which makes interpretation of the rearrangement of such an optically active system subject to much uncertainty: a mixture of antipodes in the product might imply either a lack of stereochemical specificity or a competitive operation of both fully stereoselective rearrangement pathways.

Doering and coworkers [39] have avoided this difficulty using an *optically active and deuterium labeled diester* to obtain stereochemical results fully consonant with our model. The levorotatory isomer was converted to a mixture of products from which the levorotatory antipode was obtained by resolution and found by nmr spectroscopy to have the deuterium atoms at the position shown.

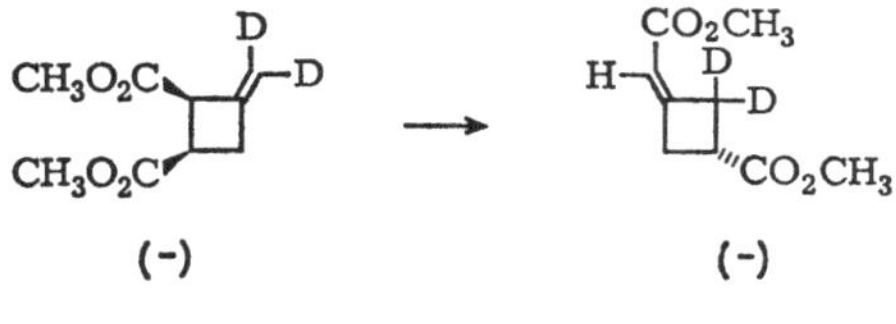

287

We expect that when the (+) isomer of the product is examined, it will have the stereochemistry shown, and that rearrangement of the stereochemically specifically *anti* monodeuterated diester would give *cis*-(-) and *trans*-(+) products.

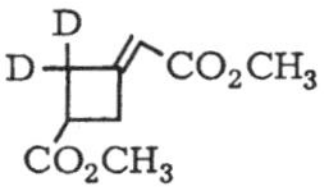

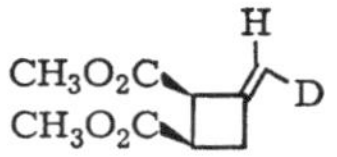

(+)	anti -(-)	cis -(-)	trans -(+)

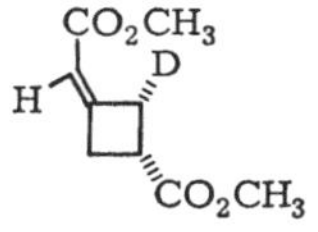

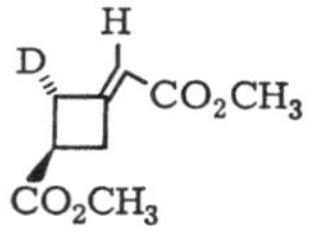

IV. Allene-Allene Cycloadditions

Allene dimerizes to give both 1,2-dimethylenecyclobutane and 1,3-dimethylenecyclobutane, as determined by the early researches of Lebedev and confirmed in the past decade by Weinstein and Fenselau [139] and by Slobodin and Khitrov [121,122]. The 1,3-dimethylenecyclobutane type product has not been reported in dimerizations involving substituted allenes, and nothing is known concerning its stereochemistry. In the following discussion, reactions giving 1,2-dimethylenecyclobutanes are considered; the interesting questions concerning what factors govern competition between the two sorts of (2+2) cycloadditions are reserved for discussion elsewhere.

Many dimerizations have been reported for allenes bearing four identical substituents: allene itself and tetramethyl [127], tetrafluoro [13,114], tetrachloro [49,80,94,99-101,103,108,128] and tetrabromo [103,106] analogs all give 1,2-dimethylenecyclobutanes completely devoid of information regarding reaction stereochemistry. Tetrachloroallene dimerizes with activation parameters $\Delta H = 11$ kcal/mole, $\Delta S = -45$ eu/mole [99].

Monosubstituted or 1,1-disubstituted allenes are the simplest cases where at least some stereochemical information on dimerizations may be gained.

Tribromoallene gives 1,2-bis(dibromomethylene)-3,4-*trans*-dibromocyclobutane [102,106]; chloroallene gives the *syn,syn* isomer of 1,2-bis(chloromethylene)cyclobutane [65]; trichloroallene gives 1,2-bis(dichloromethylene)-3,4-*trans*-dichlorocyclobutane as well as an isomer or isomers of the 1,2-bis(chloromethylene) compound [104]. Carboethoxytrichloroallene dimerizes to give 1,2-bis(carboethoxychloromethylene)-tetrachlorocyclobutane of undetermined stereochemistry [75,105].

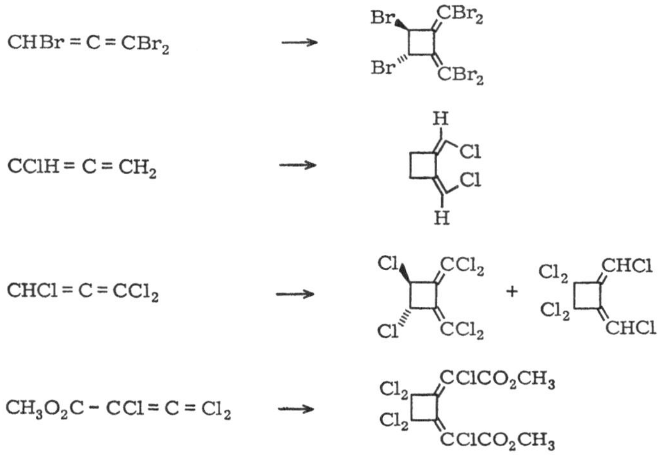

CHBr = C = CBr2
Br
CBr2
Br
CBr2
CClH = C = CH2
H
Cl
Cl
H
CHCl = C = CCl2
Cl
CCl2
Cl
CCl2
+
Cl2
CHCl
Cl2
CHCl
CH3O2C - CCl = C = Cl2
Cl2
CClCO2CH3
Cl2
CClCO2CH3

Dehmlow [38] has found that *trimethylallene* leads to a very complex mixture of dimers and triphenylallene gives both *cis* (16%) and *trans* (84%) 1,2-bis(diphenylmethylene)-3,4-diphenylcyclobutane. Triphenylallene and chlorotriphenylallene give completely different sorts of dimeric products under acid catalyzed conditions [64,67,77,78].

Dimerization of methylallene gives seven non-geminal dimethyl-1,2-dimethylenecyclobutanes at 170 °C [53].

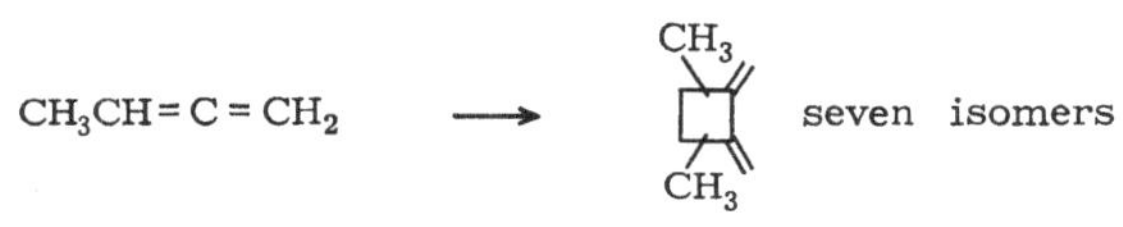

CH3CH = C = CH2
CH3
CH3
seven isomers

The rate sequence for dimerization of substituted trichloroallenes is $CN > CO_2C_2H_5 > Cl > C_6H_5 = Br > H$ [99]; products have not been reported.

1,1-Diphenylallene affords a dimer characterized as 1-diphenylmethylene-2-methylene-3,3-diphenylcyclobutane [18]. 1,1-Difluoro-3,3-bis(trifluoromethyl)allene gives at 90 °C a mixture of at least seven compounds thought to be dimers and trimers [11]. 1,1-Difluoroallene dimerizes to both the 3,3,4,4-tetrafluoro and 1-difluoromethylene-2-methylene-3,3-difluorocyclobutane [26]. Perfluoro-1,2-pentadiene reacts at 20 °C to produce three dimers, including 22% of a 1,2-bis(difluoromethylene)cyclobutane and 63% of a 1-fluoroperfluoroethylmethylene-2-difluoromethylene cycloadduct [10]. 1,1-Dichloro-3,3-dimethylallene gives a mixture of dimers [25].

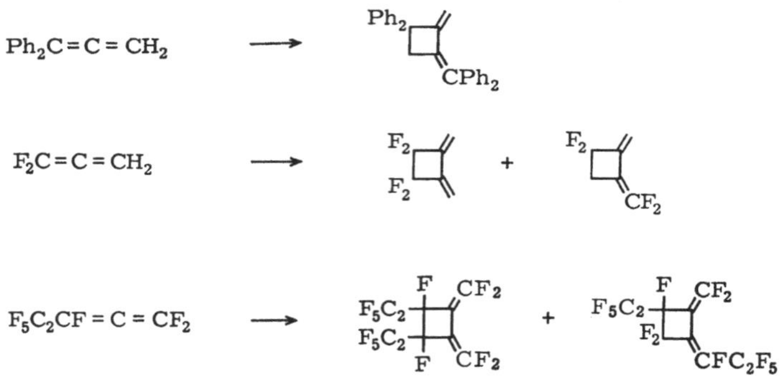

$Ph_2C=C=CH_2 \longrightarrow$

$F_2C=C=CH_2 \longrightarrow$

$F_5C_2CF=C=CF_2 \longrightarrow$

Allenes with substitution pattern A_2XY, though more complicated than the systems reviewed above, still avoid the multiplicity of options available to optically active propadienes.

1,1-Diphenyl-3-chloroallene gives a 1,2-bis(diphenylmethylene)-3,4-dichlorocyclobutane [45,77,78,86] originally assigned the *cis* stereochemistry through dipole measurements and predictions [78], but recently correctly identified as the *trans* isomer by Dehmlow [38]. A 1-chloromethylene-2-diphenylmethylene dimer is also formed [45]. 1,1-Diphenyl-3-bromoallene dimerizes to produce 1,2-bis(diphenylmethylene)-3,4-*trans*-dibromocyclobutane [38] and a 1-bromomethylene-2-diphenylmethylene adduct [82]. 1,1-Diphenyl-3-iodoallene [76] has apparently not been dimerized.

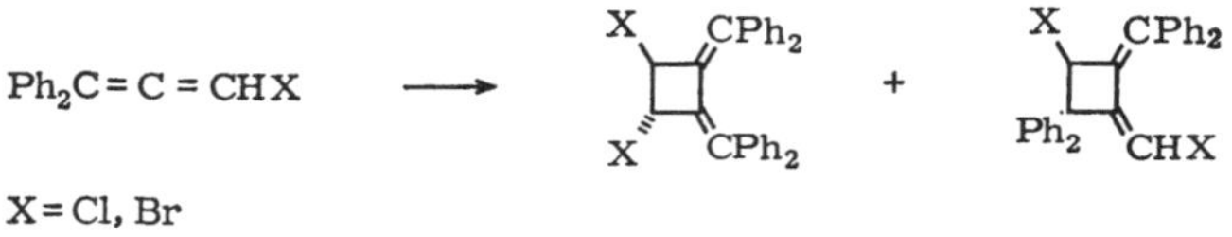

$Ph_2C=C=CHX \longrightarrow$

$X=Cl, Br$

1,1-Dimethyl-3-chloroallene gives a liquid 1-chloromethylene-2-isopropylidene and a solid 1,2-bis(isopropylidene) dimer, a mixture of *cis* and *trans* isomers according to Bertrand [24,25] and assigned the *trans* configuration by Jacobs, McClenon, and Muscio [65]. The major dimer from 1,1-dimethyl-3-bromoallene is 1,2-bis(isopropylidene)-3,4-*trans*-dibromocyclobutane [66].

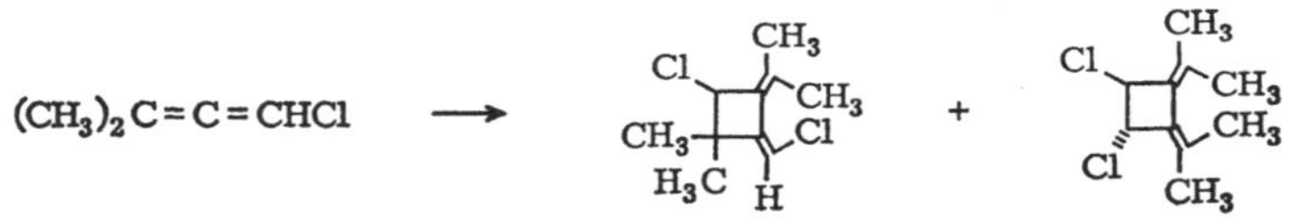

$(CH_3)_2C=C=CHCl \longrightarrow$

1,1-Diphenyl-3-diphenylmethylallene gives a 1-diphenylmethylene-3,3-diphenyl dimer under kinetic control and a 1,2-bis(diphenylmethylene) dimer as the thermodynamically favored product [119].

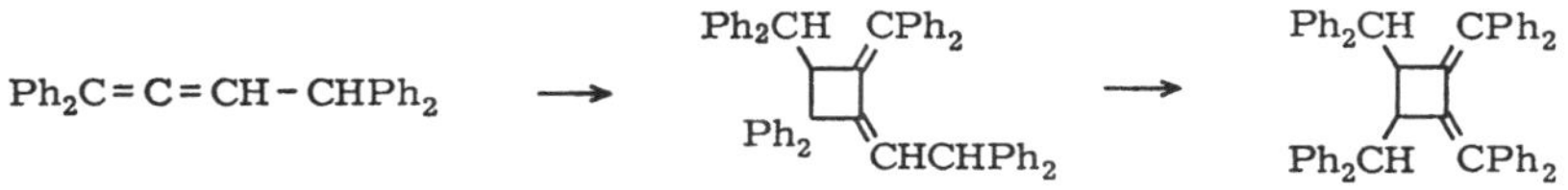

A dimer from 1,1-dichloro-3-cyano-3-methoxyallene has been reported [107].

Symmetrically disubstituted allenes, which may be optically active, have been dimerized. Racemic or optically active 1,3-diphenylallene gives both a *cis* [37,124] and a racemic *trans* [38] isomer of 1,2-bis(phenylmethylene)-3,4-diphenylcyclobutane. Racemic 1,3-dimethylallene affords a complex mixture of dimeric products [38,51], and partially resolved 1,3-dimethylallene gives a similar spectrum of dimers [51].

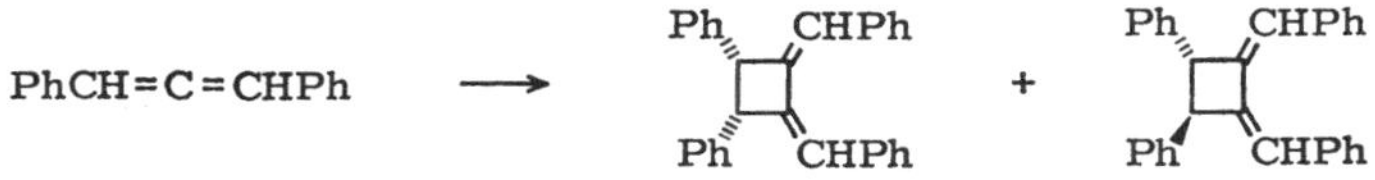

1,2-Cycloheptadiene and *1,2-cyclooctadiene* give *anti, anti* adducts [9] to which we may tentatively assign *trans* stereochemistry; both react with maleic anhydride to form 2:1 adducts, a result most plausibly interpreted in terms of a thermally allowed valence isomerization of the initially formed 1:1 adduct.

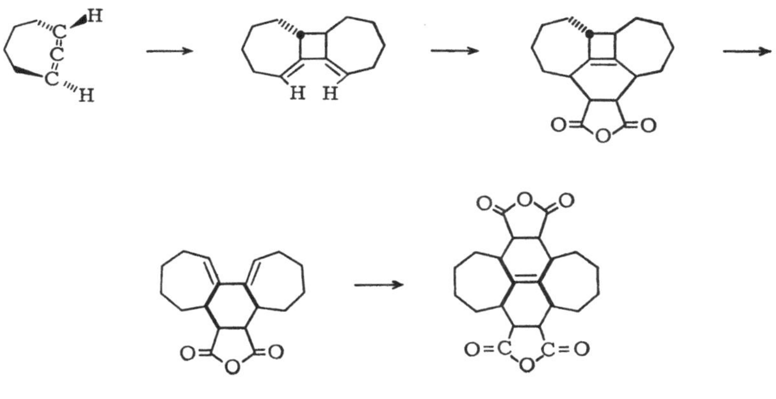

1,2-Cyclononadiene has been shown, in a most elegant set of experiments by Moore, Bach, and Ozretich [83], to dimerize with a high degree of stereoselectivity. Racemic 1,2-cyclononadiene at **125** °C gives three isomeric dimers, two of which were detected earlier by Skattebøl and Solomon [120].

from dl 6.3%	62.5%	31.5%
from d (>90%) 0.4	11.9	88.1

Dimerization of optically active 1,2-cyclononadiene gives the same three products in different proportions. At least for the most part and perhaps exclusively two allenes of the same chirality give the third, *meso* dimer, while two antipodes give rise to the first two, *dl* products.

These results are compatible with stereochemical predictions derived through orbital symmetry theory, assuming a one-step $\pi^2s + \pi^2a$ addition. But secondary deuterium kinetic isotope effects on the allene plus allene thermal (2+2) cycloaddition seem to require a two-step mechanism with formation of an intermediate [44], and as Moore and coworkers fully realized [83] stereoselective formation and reactions of 2,2'-biallylene intermediates will equally well account for the product ratios. In their rationale, two allenes approach and distort through simultaneous conrotatory twistings to give the perpendicular 2,2'-biallylene intermediate, which closes to form products in a disrotatory fashion. The experimentally observed stereochemical selectivity is equally compatible with a reversed order of rotatory motions: disrotatory joining of two allenic reactants followed by conrotatory closure to create the 1,2-dimethylenecyclobutane products [83].

These results with 1,2-cyclononadiene make possible reassessment of some other stereochemical results. In reactions of racemic allenes or partially resolved allenes [38,51], when will the rates of $R + S$ reactions be significantly different from $R + R$ or $S + S$ dimerizations, and how will these rate differences influence the observed product distributions? Even for optically inactive allenes, a serious uncertainty remains: when many products are obtained, does this result from intermediates which lose their stereochemical bearings, or may it not be a consequence of many completely stereoselective paths being kinetically competitive? Methylallene, for instance, has ten discrete modes of approach; as two of these

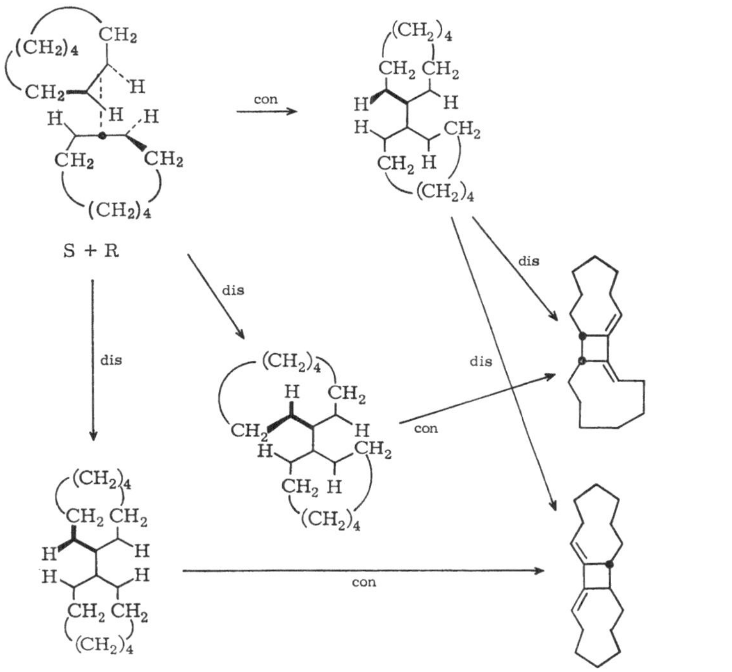
S + R
con
dis
dis
dis
dis
con
con

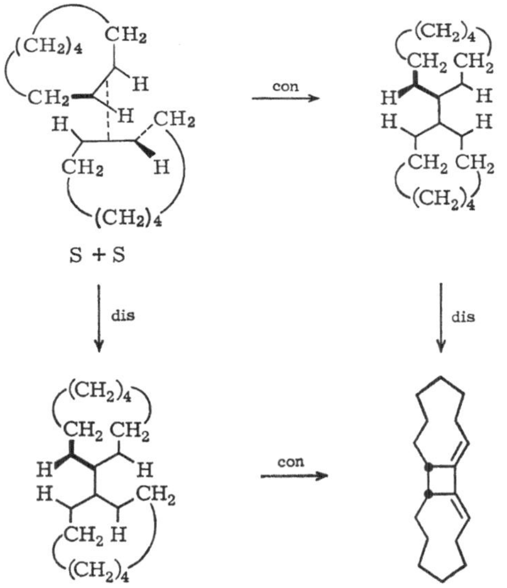
S + S
con
dis
dis
con

molecules approach, the methyl groups before the first rotatory process begins may be located in ten distinct ways. Some of these have options for the rotation step, and some of the 2,2'-biallylene intermediates may close to form the 1,2-dimethylenecyclobutane products in several ways.

1-Chloro-3-ethyl-3-methylallene and 1-chloro-3-isopropyl-3-methylallene have given at least two dimers apiece [25]. 1-Chloro-3-mesitylallene gives four dimers [85]; one is a 1,2-bis(mesitylmethylene)-3,4-dichlorocyclobutane having mesityl groups *syn* to one another. With 1-chloro-3-(1-adamantyl)allene, all three 1,2-bis(chloromethylene)-3,4-*trans*-di(1-adamantyl)cyclobutanes are formed: 28% *syn, syn*, 15% *syn, anti*, and 13% *anti, anti* [63]. Three other dimers have been isolated and partially clarified structurally.

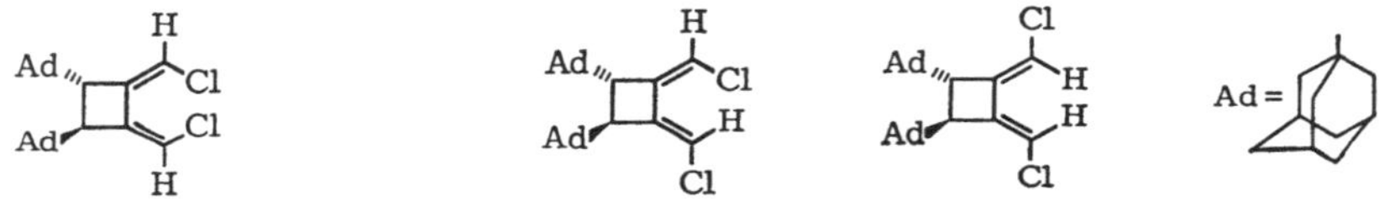

1,3-Diphenylallene and 1,1,3-triphenylallene give mixed dimers with tetrachloroallene [36].

V. 1,2-Dimethylenecyclobutane Rearrangements

Gajewski and Shih [52] synthesized 1,2-bis(dideuteriomethylene)cyclobutane and found it rearranged at 275 °C to a mixture of tetradeuterio-1,2-dimethylenecyclobutanes containing 26—31% of the total protium on exocyclic methylene groups.

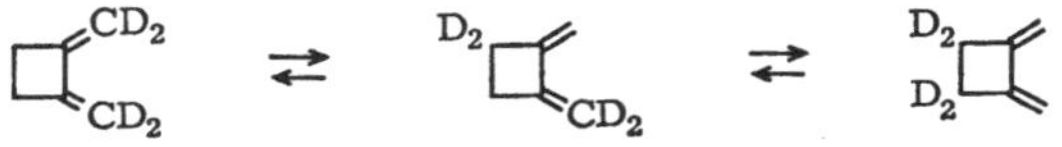

The reaction was shown to be intramolecular and estimated to have an activation energy of 43.5 kcal. Allenylcyclobutane, a conceivable intermediate for the deuterium scrambling, did not rearrange to 1,2-dimethyl-enecyclobutane to any significant extent under the reaction conditions. The isomerized tetradeuterio material and dimethyl acetylenedicarboxy-late gave a Diels-Alder adduct, which was dehydrogenated to afford the corresponding benzocyclobutene, 58% d_2, 28% d_3, and 14% d_4.

This result suggests that 1-dideuteriomethylene-2-methylene-3,3-dideuteriocyclobutane was favored kinetically by a factor of 2 over 1,2-dimethylene-3,3,4,4-tetradeuteriocyclobutane, that is, exactly as one would predict from a perpendicular 2,2'-biallylene intermediate partitioning according to the relevant statistical factors.

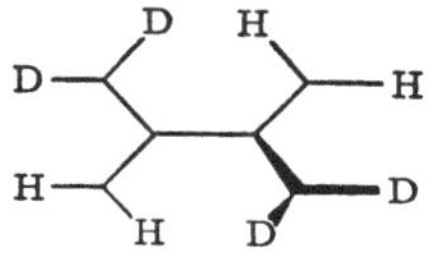

Doering and Dolbier [40] similarly observed the degenerate rearrange-ment and determined the reaction kinetics at 3—5 mm from 261.5 to 299.0 °C; the Arrhenius plot gave $E_a = 46.8 \pm 0.9$ kcal/mole and log $A = 14.45$. This result is quite surprising when justaposed with Arrhenius parameters for the degenerate methylenecyclobutane rearrangement [41], $E_a = 49.5 \pm 1$ kcal/mole and log $A = 14.77$. Although formally the 1,2-dimethylenecyclobutane rearrangement might profit from two allylic resonance energy decrements while methylenecyclobutane may from only one, they have very comparable kinetic characteristics.

The stereochemistry of the rearrangement has been the subject of a recent communication by Gajewski and Shih [53]. They determined that 1,2-dimethylene-3,4-*trans*-dimethylcyclobutane rearranged thermally al-most exclusively ($>94\%$) to the *anti* isomer of 1-ethylidene-2-methylene-3-methylcyclobutane. Starting with the 3,4-*cis*-dimethyl analog, the major products were the *syn* and *anti* isomers of 1-ethylidene-2-methylene-3-methylcyclobutane, the *syn, anti* isomer of 1,2-bis(ethylidene)cyclo-butane, and a triene derived from *syn*-1-ethylidene-2-methylene-3-methylcyclobutane. These stereochemical results were seen as evidence indicating conrotatory rather than disrotatory opening of the cyclo-butane ring [53]:

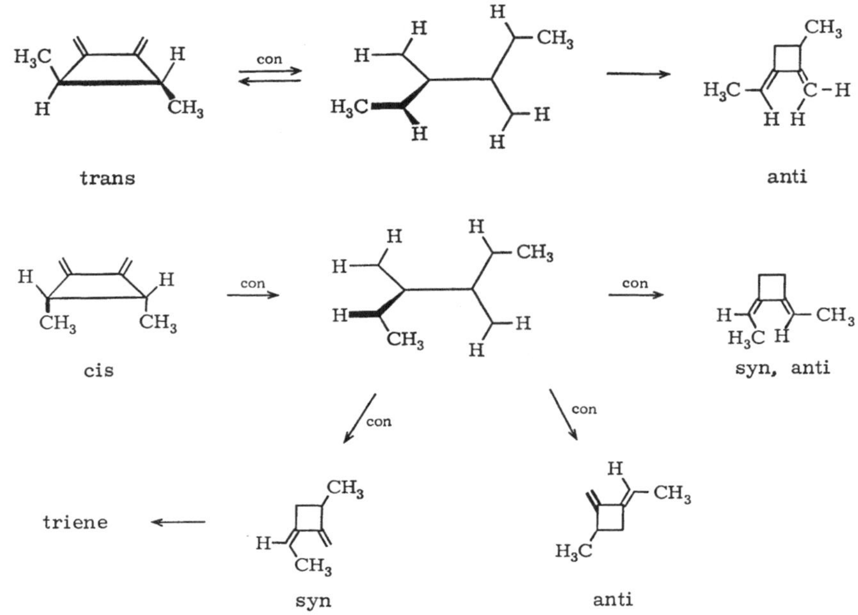

To maintain that disrotation is the stereochemical mode for ring opening, one would have to postulate that the *syn, anti* isomer is not a primary product in the thermal rearrangement of the *cis*-3,4-dimethyl system, but instead comes from a subsequent isomerization of the *syn* isomer.

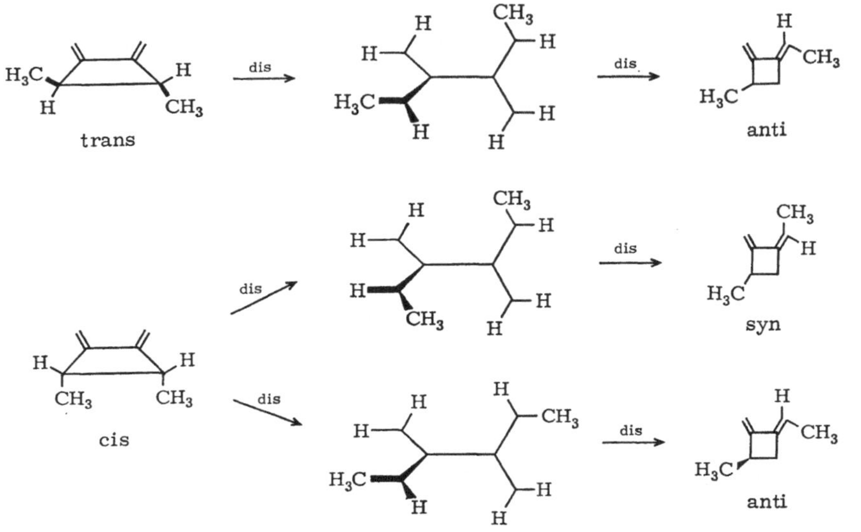

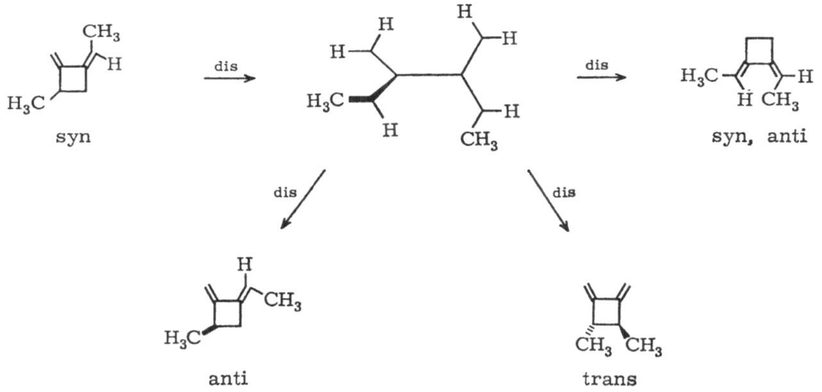

This alternative position, seemingly bolstered by the fact that the *trans* isomer is a significant thermal product when the *cis* compound is rearranged, requiring intervention of at least one other dimethyl-1,2-dimethylenecyclobutane isomer and a secondary reaction from it if the exclusive operation of either mode of ring opening be assumed, may not, however, account quantitatively for the product ratios. The disrotatory model would require different partitioning ratios from the one-methyl-in, one-methyl-out biallylene intermediate from the *trans* compound, and another from the *syn*. This contradiction and impossibility, not only the parsimonious assumption that secondary reactions may be neglected, forces the conclusion: in 1,2-dimethylenecyclobutane rearrangements, cyclobutane rings are opened and made again in a conrotatory fashion.

In analogy with the spiropentane to methylenecyclobutane rearrangement, one may expect that kinetic studies on 1-methylenespiropentane will be on interest. Thermal reactions of substituted methylenespiropentane have just been reported [33].

VI. Molecular Orbital Calculations

Theoretical analysis of the methylenecyclobutane and 1,2-dimethylene-cyclobutane rearrangements using the extended Hückel (EH) molecular orbital method gave estimates of energy as a function of structure. We employed a modified version of the program originally written by R. Hoffmann [61], adapted to the IBM 360/50 at the University of Oregon with the helpful assistance of C. E. Klopfenstein and H. Merl. The H_{ij} terms were evaluated using the formula $H_{ij} = 0.875\, S_{ij}(H_{ii} + H_{jj})$; calculations were not iterated to secure constant charge densities except as noted. Other parameters were identical to those previously employed [5,71].

The energy of methylenecyclobutane *(1)* calculated for an idealized square planar C_{2v} geometry having $C(1)-C(2) = 1.34$ Å, other C—C bonds 1.56 Å, C—H $= 1.09$ Å, $< HC(1)H = 120°$ and other $< HCH = 116°$ was taken as a reference; other energies quoted for C_5H_8 systems below are relative to methylenecyclobutane.

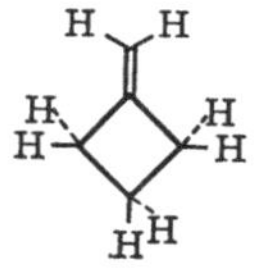

1 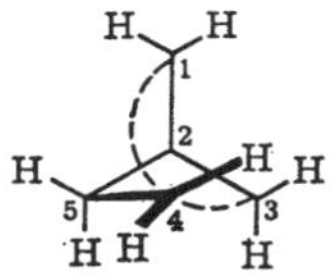   *2* 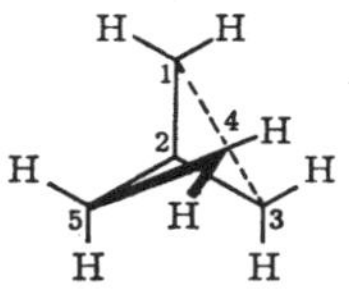   *3*

A one step mechanism for the methylenecyclobutane rearrangement would presumably go by a 1,3-sigmatropic migration of C(4) from C(3) to C(1) with inversion of configuration at the migrating carbon [20,109]. To assess the likelihood of this process, calculations were done on models for the hypothetical transition state configuration of atoms. With $C(1)-C(2) = C(2)-C(3) = 1.45$ Å, $C(3)-C(4) = C(1)-C(4) = 1.67$ Å, $C(2)-C(5) = C(5)-C(4) = 1.56$ Å, $< C(1)C(2)C(3) = < C(1)C(2)C(5) = < C(3)C(2)C(5) = 90°$ and $< HCH = 116°$, the structure *2* of C_s symmetry was found to be 238 kcal/mole higher in energy than methylenecyclobutane.

An alternative geometry *3*, with C—C $= 1.56$ Å, $< HCH = 120°$, and $< C(2)C(5)C(4) = 75°$; the atoms except C(4) and attached hydrogens having C_{3v} symmetry; and C(4) displaced 0.1 Å from the C_3 axis away from C(5), had a calculated energy of 217 kcal/mole higher than the reference.

Using iterative calculations to reach constant charge densities, structures *1*, *2*, and *3* were found to have relative energies of 0, $+361$ and $+376$ kcal/mole respectively. Orbital symmetry requirements for concerted processes are met in the activated complex, but the energetic disabilities associated with the contorted goemetry are prohibitive. The structures *2* and *3* are inaccessibly higher in energy than methylenecyclobutane, and one may conclude that the concerted migration-with-inversion mechanism is unlikely.

Favored configurations for the 2-(dimethylene)allylene system *4* were sought through calculations in which the dihedral angles α for HC(3)C(2)C(5), β for C(4)C(5)C(2)C(3), and γ for HC(4)C(5)C(2) were systematically varied. The geometrical parameters employed were

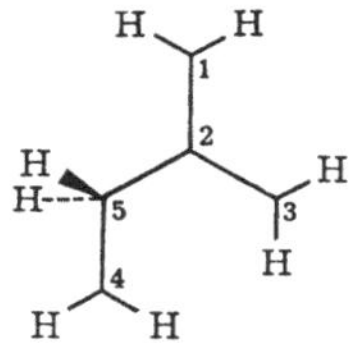

4 (0, 0, 0) 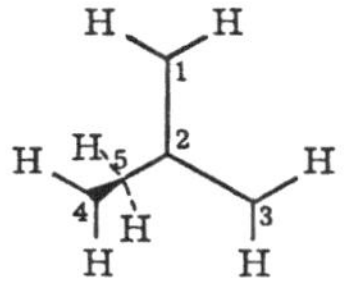 4 (0, 90, 90) 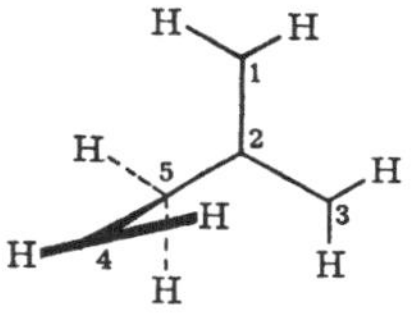 4 (0, 90, 0)

Dihedral angles in 4:

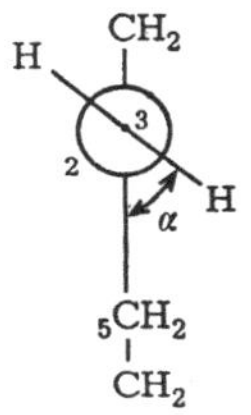

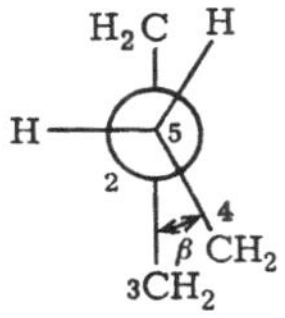

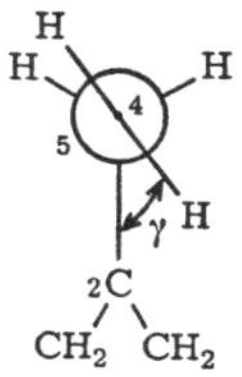

$C(1)-C(2) = C(2)-C(3) = 1.54$ Å, $C(5)-C(2) = 1.55$ Å, $C(4)-C(5) = 1.50$ Å, regular trigonal $120°$ bond angles except $<C(2)C(5)C(4) = 105°$ and $<HC(5)H = 114°$. The calculated total energies relative to methylenecyclobutane are listed in Table 1.

Table 1. *Calculated energies for conformation isomers of 2-(dimethylene)allylene 4*

Dihedral Angles (°) in 4			Energy*) (kcal/mole)
α	β	γ	
0	0	0	33.1
0	0	45	20.0
0	0	90	11.6
0	45	0	13.3
0	45	45	12.6
0	45	90	7.5
0	90	0	7.6
0	90	45	6.7
0	90	90	6.2
90	0	0	44.7
90	0	90	29.2

*) Relative to methylenecyclobutane.

According to these results, the planar geometry for species 4 $(\alpha, \beta, \gamma) =$ (0, 0, 0) is of comparatively high energy, and of the two geometries in which the C(2)C(5)C(4) plane is perpendicular to the C(1)C(2)C(3) plane $(\beta = 90)$, there is an apparent preference for the hydrogens on C(4) to be perpendicular to the C(2)C(5)C(4) plane $(\gamma = 90)$ rather than in that plane $(\gamma = 0)$. Further, changes in any of the three angular variables from the values (0, 90, 90) are destabilizing, suggesting that this (0, 90, 90)-form corresponds to a local energy minimum, relative to other models in this highly restricted set.

The approximations inherent in the EH method make the small energy preference calculated for 4 (0, 90, 90) over 4 (0, 90, 0) seem inadequate grounds for firm conclusions. We simply note that the preference for 4 (0, 90, 90) as a stereochemical model for the intermediate in allene-olefin cycloadditions deduced from experimental data is consonant with the EH result. The very small dependence in energy upon the angle γ for 4 (0, 90, γ) may be cause for suspecting that at least in some substituted cases, rotation through $\gamma = 180\,^\circ$ in the intermediate 4 (0, 90, γ) may be kinetically competitive with ring closure.

For the 2,2'-biallylene intermediate, with C(2)C(5) = 1.55 Å, other CC bonds 1.45 Å, and idealized trigonal geometry, we find the planar D_{2h} form 13.8 kcal/mole higher in energy than the perpendicular D_{2d} intermediate. Hoffmann has calculated an energy difference of 4 kcal/mole [53] (note 7), in favor of the perpendicular form, and indicated that SCF π-electron calculations give nearly identical energies for the two species.

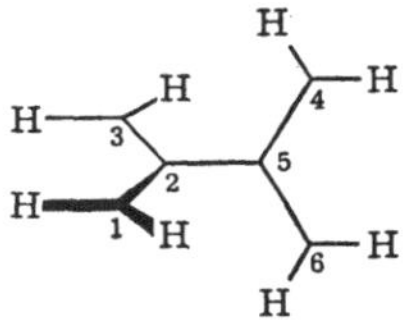

D_{2d} 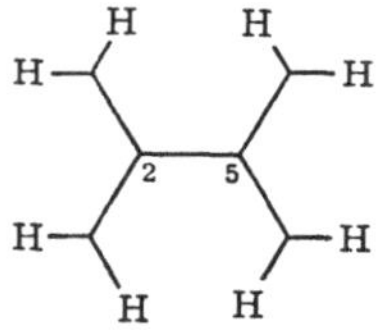 D_{2h}

VII. Concertedness

An analysis and interpretation of experimental results for the four reactions here considered in terms of the Woodward-Hoffmann theory for *concerted reactions* must be predicated on a clear and commonly held understanding of what that term may mean. Despite the centrality of this term and the concept it covers in recent theoretical treatments of cyclo-addition and more general cycloreaction mechanisms, it seems to us to have been and still to be a source of some confusion. The confusion may be traced to three distinct conceptions of concertedness, often used loosely as meaning more or less the same thing. We will attempt to disentangle these conceptions and see how they relate to the problems at issue.

Concertedness in a chemical process is seen by some in terms of an energy surface in configuration space. An *energetically concerted* process has no local potential minimum, sufficiently deep to allow reactants to reside there for at least a few vibrational periods, separating the energy minima characteristic of reactants and products. It has a reaction profile with no potential energy wells between reactant and product deeper than the minimum $1/2\ h\nu_0$ vibrational energy value of the reacting system at that point in configuration space. Thus an energy profile characterizes a total process as concerted or not, and a precise definition of how large an irregularity in that profile must be before it may become the potential well of a reactive intermediate is available. The curves in Figure 1 illustrate the point.

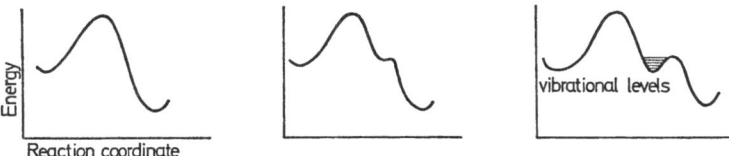

Fig. 1. Energy versus reaction coordinate profiles for two concerted and a nonconcerted process

One conceivable extrapolation of this definition of concertedness is that every elementary chemical reaction is concerted. Every chemical process may be at least conceptually broken up into a sequence of one, two, or more elementary steps or concerted subprocesses. Therefore a reaction may be termed one-step, two-step, or multi-step, and the term "concerted" becomes a redundancy simply equivalent to "one-step". The frequent use of "concerted" and "one-step" as interchangeable adjectives is one consequence of this extrapolation.

A second view of concertedness has been succinctly put by Bartlett [15]: in *concerted cycloadditions*, bonding is established simultaneously at more than one site. Individually discrete aspects of an overall concerted process are coupled together synchronously or simultaneously so that at no point along the reaction path must the system pay the high energetic price demanded for less well organized, piecemeal approaches toward the product. The diagrammatic hallmark of this description would be plots of bond indices versus reaction coordinates, comparing the progress of two or more subevents or aspects of an overall reaction (Fig. 2).

In *bondingly concerted reactions*, two or more new bonds form synchronously, at exactly the same time and rate, or simultaneously, at the same time.

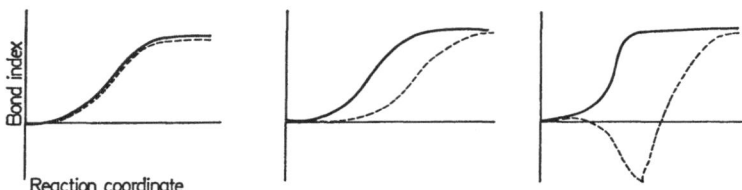

Fig. 2. Plots of bond indices for two new bonds formed in a cycloaddition for synchronous and simultaneous concerted reactions, and for a nonconcerted process

As Fig. 2 shows at the right, nonconcerted reactions force one new bonding interaction to take place at the expense of another new antibonding one. The second bond may increase in strength only after the first bond is made.

A third standard for concertedness may be based on molecular orbital theory [62]. A finite band of doubly occupied molecular orbitals defines the electronic structure of an even-electron neutral ground state hydrocarbon. A set of unoccupied antibonding molecular orbitals reside at much higher energy. Some reactions, designated as concerted or symmetry allowed by Woodward and Hoffmann, may continuously transform the occupied, bonding molecular orbitals or reactant to those of product, preserving the bonding character of these orbitals throughout the transformation. Other nonconcerted reactions can't manage such a transformation: the molecular orbital situation at a mid-point of the reaction has two nonbonding orbitals which must accommodate two electrons previously associated with a bonding orbital in starting material. The reaction has a correspondingly high activation energy, and the mid-point structure is a singlet or triplet diradical intermediate. The distinguishing feature of an *orbitally concerted reaction* thus may be seen in a molecular orbital energy-level diagram (Fig. 3); throughout the reaction bonding orbitals remain bonding. In a nonconcerted process, two nonbonding orbitals and the corresponding type of high energy intermediate characterize the mid-reaction situation.

The triple-vision-of-concertedness problem may be corrected by noting that there is no inconsistency between a reaction being concerted according to bond order or molecular orbital criteria while being nonconcerted energetically. Even if two or more new bonds are formed simultaneously, and if bonding molecular orbitals are available for all electrons throughout a reaction, there still may be a local minimum in potential energy at some point along the reaction coordinate.

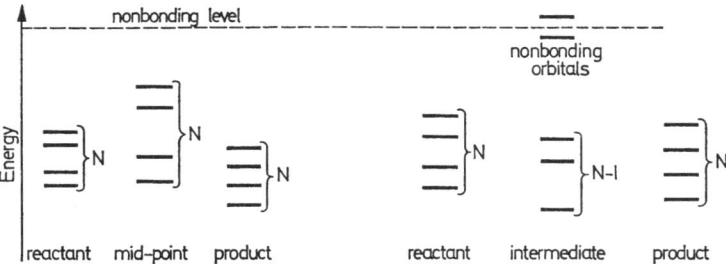

Fig. 3. Schematic molecular orbital level diagrams for concerted and nonconcerted process involving N bonding orbitals in the reactant and product

Salem [110,111)] as well as Bartlett and Schneller [15)] have approached this point, and written of "retarded one-step" and "thoroughly concerted two-barrier" reactions. We find it meaningful to describe such processes as bondingly and orbitally concerted, yet energetically nonconcerted reactions, or more tersely as concerted, two-step reactions. Thus every energetically concerted reaction must be concerted in the other senses, but not vice versa.

All electrons in the molecular intermediates involved in concerted two-step reactions may reside in bonding orbitals. They are unstable relative to our normal frames of reference, yet stable relative to small displacements in internuclear coordinates. They are quite unlike the intermediates characteristic of nonconcerted processes having two electrons in two non-bonded orbitals. They represent a class of intermediates but little studied in intramolecular reactions, and having unknown intermolecular dynamic possibilities and spectroscopic properties. These intermediates would have high energy because structurally they depart from geometric norms common to low energy, easily isolable organic molecules. In them non-integral bond indices may be frequently encountered.

VIII. Discussion and Conclusions

Kinetic isotope effects indicate that both allene-olefin and allene-allene cycloadditions occur in at least two steps, and stereochemical results suggest they may be highly stereoselective. These facts and analysis of the categories included under the rubric of concertedness leads to rationalization of these $(2+2)$ cycloadditions as orbitally concerted but energetically nonconcerted, or equivalently, as concerted two-step processes.

For the allene-olefin system, orbital symmetry controls the stereochemistry leading to the 2-(dimethylene)allylene intermediate as though the reaction were to be an energetically concerted $\pi^2s + \pi^2s + \pi^2s$ process, but the system apparently finds a local potential energy minimum and is thereafter able to bond at either end of the allyl unit. Independent evidence for the methylenecyclobutane rearrangement [39] supports the stereochemical model developed [8].

For the 1,2-dimethylenecyclobutane rearrangement, Gajewski and Shih [53] have demonstrated preferred conrotatory ring opening and closing. For the allene-allene cycloaddition, then, the sequence would be disrotatory motion as two allenes approach to form the perpendicular biallylene intermediate, followed by conrotatory closure of that species.

How may the disrotatory combination of two allenes be explained?

If all orbitals were to be involved in a concerted process, the allene thermal dimerization would be classified as a $\pi^2s + \pi^2s + \pi^2s + \pi^2a$ cycloaddition. This geometrical designation of orbitals implies a disrotatory interaction as the allenes combine.

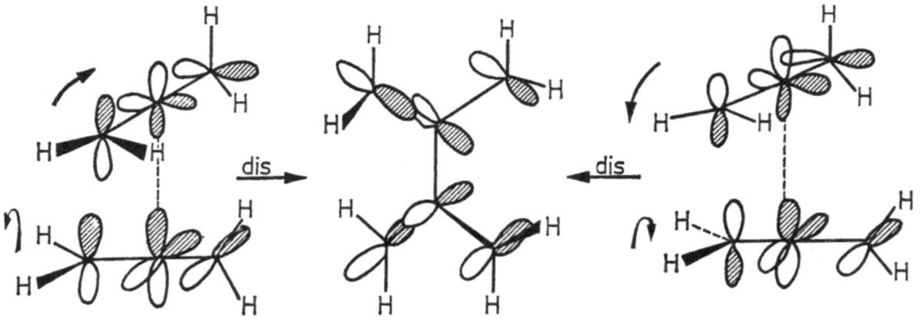

On a strict symmetry basis, conversion of 1,2-dimethylenecyclobutane to the planar biallylene with preservation of a plane of symmetry is thermally allowed with a disrotatory opening [85]. Direct conversion to the perpendicular D_{2d} species maintained a C_2 axis gives a thermally allowed conrotatory process. Thus knowing the the preferred stereo-chemistry allows construction of the relevant correlation diagram.

A recent X-ray single crystal structure determination for the dimer of tetrachloro-1,2,3-butatriene, perchloro[4]radialene [57], shows it to have a highly puckered D_{2d} molecular symmetry rather than the planar D_{4h} alternative [130]. The dihedral angle for the four-membered ring is 153.5 ± 0.6 °.

On the basis of this result [130], experience accumulated while attempt-ing to calculate dipole moments for 1,2-dimethylene -3,4-dihalocyclo-butanes assuming a planar carbon skeleton [38], and the relative ease with which bulky substituents on exocyclic carbons in 1,2-dimethylenecyclo-butanes are accommodated in *syn, syn* configurations [63,65,85], we expect representative 1,2-dimethylenecyclobutanes to have the nonplanar C_2 symmetry, rather than planar C_{2v} molecular geometry. This molecular geometry is, of course, completely accordant with the motion of nuclei required for rearrangement to the perpendicular 2,2'-biallylene in a conro-tatory fashion. Exaggeration of the nonplanarity, together perhaps with a rocking motion of the methylene groups on C(2) and C(3) to increase the staggering of vicinal hydrogens [143], leads into the conrotatory ring-opening event.

Methylenecyclobutane, having a small barrier to ring inversion in a double minimum potential function for ring puckering and relatively easy access to highly puckered geometries as shown by the microwave spectroscopic studies of Scharpen and Laurie [112], may also be converted to the 2-(dimethylene)allylene intermediate through an analogous dis-rotatory molecular distortion.

Allene and ketene, though isoelectronic, have quite different photo-electron spectra and electronic structures [4]; that they undergo $(2 + 2)$ cycloadditions with olefins by quite distinct mechanisms [6,8] is not alto-gether surprising.

Recent structural and spectroscopic investigations of organometallic complexes bonding two carbons of an allenic ligand to one rhodium [59,72, 87,95] or platinum atom [58,87,98,132] may have some pertinence to possible bridged intermediates proposed for various electrophilic additions to allenes, and the σ-iron-π-iron complexes derived from allene and diiron enneacarbonyl by Ben-Shoshan and Pettit [19] and the subject of an X-ray crystallographic investigation by Davis [35] seem to be geometrically analogous to the postulated 2-(dimethylene)allylene structure for the intermediate in allene-olefin cycloadditions.

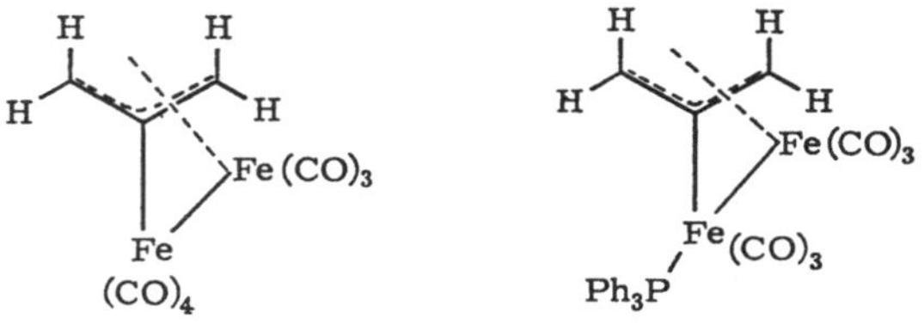

In the crystalline triphenylphosphine derivative [35] both iron atoms are on the same side of the plane defined by the allenic carbons. The stereoscopic drawing of the complex reproduced in the communication is particularly revealing of geometrical character [35]. Unfortunately, no evidence could be obtained regarding the angle between an HCH plane and the CCC plane; they certainly need not be identical [56]. One hopes that similar complexes of tetramethylallene will soon be synthesized and structurally clarified.

The molecular orbital calculations for the 2-(dimethylene)allylene system discussed above did not give a minimum energy through varying many geometrical parameters; thus the optimum geometry for 4 $(0, 90, 90)$ is still very much an open question.

The complete omission of specific material on allene cycloadditions in the Woodward and Hoffmann review article [142] emphasizes the ticklish theoretical situation; there may be no violations to the principle of maximum bonding [142], but sure applications of the theory in its present early stage of development are more easily made for some types of reaction than for others.

Rather than posing the question "What controls stereochemistry in a 'nonconcerted' reaction where a 'nonbonded' state intervenes?" we favor framing the problem according to a different perspective: In a stereochemically well defined set of interlinked events suggestive of some orbitally concerted chemical process, but where zeroth order molecular orbital theory and some doubtful assumptions concerning geometrical parameters lead to a prediction that the reaction ought to be orbitally nonconcerted, what lines of evidence can be brought to bear in searches for intermediates indicative of an energetically nonconcerted process, and what reformulations of molecular orbital analysis will produce an understanding of how comparatively unstable ground state molecules minimize total energy when the usual solutions to the bonding problem are inaccessiable for geometrical reasons? What structural circumstances in systems allowed by simple theory to react in an orbitally concerted fashion dictate whether they will occur in an energetically concerted or energetically nonconcerted fashion?

The "nonbonded" states anticipated by simple theory may in some cases be avoided by a system able to reach a "bonded" state of high energy, relative to a normal frame of reference defined by the heats of formation of indefinitely stable molecules. Such high energy states would be "bonded" if they had all electrons in bonding molecular orbitals, the highest being well removed from the nonbonding energy level. It will doubtless be argued whether intermediates of the class under consideration, such as 2-(dimethylene)allylene and 2,2'-biallylene, may live long enough to participate in intramolecular reactions, such as trapping events, or to be observed spectroscopically — at least until such experiments are successfully accomplished.

We may anticipate sustained experimental and theoretical activity on allene cycloadditions and on the degenerate rearrangements of derived cycloadducts both for intrinsic reasons and for the importance these areas may play in providing a testing ground for new conceptual hypotheses. Many stereochemical features of known allene-allene dimerizations remain unsettled, and new systems will increasingly be designed, synthesized, and investigated in response to and as a needed corrective of current understandings and speculations.

Acknowledgment. For the financial support of our current research on cycloreactions of hydrocarbons we thank Cities Service Oil Company, the duPont Company, the Petroleum Research Fund, and the Research Corporation.

IX. References

[1] Andrews, S. D., Day, A. C.: Chem. Commun. 902 (1967).
[2] Bach, R. D.: Tetrahedron Letters 5841 (1968).
[3] — J. Am. Chem. Soc. *91*, 1771 (1969).
[4] Baker, O., Turner, D. W.: Chem. Commun. 480 (1969).
[5] Baldwin, J. E., Foglesong, W. D.: J. Am. Chem. Soc. *90*, 4311 (1968).
[6] — Kapecki, J. A.: J. Am. Chem. Soc. *91*, 3106 (1969).
[7] — — Roy, U. V.: unpublished.
[8] — Roy, U. V.: Chem. Commun. 1225 (1969).
[9] Ball, W. J., Landor, S. R.: J. Chem. Soc. 2298 (1962).
[10] Banks, R. E., Braithwaite, A., Haszeldine, R. N., Taylor, R. E.: J. Chem. Soc. C 2593 (1968).
[11] — — — — J. Chem. Soc. C 996 (1969).
[12] — Deem, W. R., Haszeldine, R. N., Taylor, D. R.: J. Chem. Soc. C 2051 (1966).
[13] — Haszeldine, R. N., Taylor, D. R.: J. Chem. Soc. 978 (1965).
[14] — — — J. Chem. Soc. 5602 (1965).

15) Bartlett, P. D., Schneller, K. E.: J. Am. Chem. Soc. 90, 6077 (1968).
16) — Science 159, 833 (1968).
17) Battioni, P., Aspect, A., Vo Quang, L.: Compt. Rend., Ser. C 268, 1263 (1969).
— Battioni, P., Vo Quang, Y.: Compt. Rend. 266, 1310 (1968).
18) Beltrame, P., Pitea, D., Marzo, A., Simonetta, M.: J. Chem. Soc. B 71 (1967).
19) Ben-Shoshan, R., Pettit, R.: Chem. Commun. 247 (1968).
20) Berson, J. A., Nelson, C. L : J. Am. Chem. Soc. 89, 5503 (1967).
21) Bertrand, M., Grimaldi, J., Waegell, B.: Chem. Commun. 1141 (1968).
22) — Le Gras, J.: Bull. Soc. Chim. France, 4336 (1967).
23) — — Bull. Soc. Chim. France 3044 (1968).
24) — Maurin, R.: Compt. Rend., Ser. C 265, 609 (1967).
25) — Reggio, H., Leandri, G.: Compt. Rend. 259, 827 (1964).
26) Blomquist, A. T., Nicholas, P. P.: J. Org. Chem. 32, 866 (1967).
27) Borden, W. T., Corey, E. J.: Tetrahedron Letters 313 (1969).
28) Brandaur, R. L., Short, B., Kellner, S. M. E.: J. Phys. Chem. 65, 2269 (1961).
29) Brewster, J. H.: Topics Stereochem. 2, 1 (1967).
30) Burkhardt, P. J.: Dissertation, University of Oregon, Eugene, Oregon (1962).
31) Chesick, J. P.: J. Phys. Chem. 65, 2170 (1961).
32) — J. Am. Chem. Soc. 85, 2720 (1963).
33) Crandall, J. K., Paulson, D. R., Bunnell, C. A.: Tetrahedron Letters 4217 (1969).
34) Crombie, L., Jenkins, P. A.: Chem. Commun. 870 (1967).
35) Davis, R. E.: Chem. Commun. 248 (1968).
36) Dehmlow, E. V.: Chem. Ber. 100, 2779 (1967).
37) — Chem. Ber. 100, 3260 (1967).
38) — Tetrahedron Letters 4283 (1969).
39) Doering, W. von E.: Lecture, Twenty-First National Organic Chemistry Symposium of the American Chemical Society, Salt Lake City, June 18, 1969.
40) — Dolbier, W. R., Jr.: J. Am. Chem. Soc. 89, 4534 (1967).
41) — Gilbert, J. C.: Tetrahedron, Suppl. 7, 397 (1966).
42) — — Leermakers, P. A.: Tetrahedron 24, 6863 (1968).
43) Dolbier, W. R., Jr., Dai, S. H.: J. Am. Chem. Soc. 90, 5028 (1968).
44) — — J. Am. Chem. Soc. 92, 1774 (1970).
45) Doupeux, H., Martinet, P.: Compt. Rend., Ser. C 262, 588 (1966).
46) Fedorova, A. V., Petrov, A. A.: Zh. Obshch. Khim. 32, 3537 (1962).
47) Flowers, M. C., Frey, H. M.: J. Chem. Soc. 5550 (1961).
48) Frey, H. M.: Trans. Faraday Soc. 57, 951 (1961).
49) Fujimo, A., Nagata, Y., Sakan, T.: Bull. Chem. Soc. Japan 38, 295 (1965).
50) Gajewski, J. J.: J. Am. Chem. Soc. 90, 7178 (1968).
51) — Black, W. A.: Tetrahedron Letters 899 (1970).
52) — Shih, C. N.: J. Am. Chem. Soc. 89, 4532 (1967).
53) — — J. Am. Chem. Soc. 91, 5900 (1969); and unpublished.
54) Griesbaum, K.: Angew. Chem. Intern. Ed. Engl. 5, 933 (1966).
55) Harris, J. F., Jr.: Tetrahedron Letters 1359 (1965).
56) Heimbach, P., Traunmüller, P.: Ann. Chem. 727, 208 (1969).
57) Heinrich, B., Roeding, A.: Angew. Chem. Intern. Ed. Engl. 7, 375 (1968).
58) Hewitt, T. G., Anzenhofer, K., DeBoer, J. J.: J. Organometal. Chem. (Amsterdam) 18, 19 (1969).
59) — — — Chem. Commun. 312 (1969).
60) Hoff, S., Brandsma, L., Arens, J. F.: Rec. Trav. Chim. 87, 1179 (1968).
61) Hoffmann, R.: J. Chem. Phys. 39, 1397 (1963).
62) — Woodward, R. B.: Accounts Chem. Res. 1, 17 (1968).

63) Jacobs, T. L.: Lecture, University of Oregon, November 14, 1969.
64) — Dankner, D., Singer, S.: Tetrahedron 20, 2177 (1964).
65) — McClenon, J. R., Muscio, O. J., Jr.: J. Am. Chem. Soc. 91, 6038 (1969).
66) — Petty, W. L.: J. Org. Chem. 28, 1361 (1963).
67) Jones, D. W.: J. Chem. Soc. C 1026 (1966).
68) Jones, F. N., Lindsey, R. V., Jr.: J. Org. Chem. 33, 3838 (1968).
69) Jones, W. M., Walbrick, J. M.: Tetrahedron Letters 5229 (1968).
70) Joshi, G. C., Devaprabhakara, D.: J. Organometal. Chem. (Amsterdam) 15, 497 (1968).
71) Kapecki, J. A., Baldwin, J. E.: J. Am. Chem. Soc. 91, 1120 (1969).
72) Kashiwagi, T., Yasuoka, N., Kasai, N., Kukudo, M.: Chem. Commun. 317 (1969).
73) Kiefer, E. F., Okamura, M. Y.: J. Am. Chem. Soc. 90, 4187 (1968).
74) — Tanna, C. H.: J. Am. Chem. Soc. 91, 4478 (1969).
75) Köbrich, G., Wagner, E.: Angew. Chem. Intern. Ed. Engl. 7, 470 (1968).
76) Kollmar, H., Fischer, H.: Tetrahedron Letters 4291 (1968).
77) Landor, P. D., Landor, S. R.: Proc. Chem. Soc. (London), 77 (1962).
78) — — J. Chem. Soc. 2707 (1963).
79) Lukina, M. Yu.: Usp. Khim. 32, 1425 (1963).
80) Maass, G.: Angew. Chem. Intern. Ed. Engl. 2, 394 (1963).
81) Marrov, M. V., Kucherov, V. F.: Usp. Khim. 36, 553 (1967).
82) Martinet, P., Doupeux, H.: Compt. Rend. 261, 2498 (1965).
83) Moore, W. R., Bach, R. D., Ozretich, T. M.: J. Am. Chem. Soc. 91, 5918 (1969).
84) Moriconi, E. J., Kelly, J. F.: J. Org. Chem. 33, 3036 (1968).
85) Muscio, O. J., Jr., Jacobs, T. L.: Tetrahedron Letters 2867 (1969).
86) Nagase, T.: Bull. Chem. Soc. Japan 34, 139 (1961).
87) Osborn, J. A.: Chem. Commun. 1231 (1968).
88) Otsuka, S., Nakamura, A., Minamida, H.: Chem. Commun. 191 (1969).
89) — — Tani, K.: J. Organometal. Chem. (Amsterdam) 14, P30 (1968).
90) — Tani, K., Nakamura, A.: J. Chem. Soc. A 1404 (1969).
91) — Nakamura, A., Tani, K., Ueda, S.: Tetrahedron Letters 297 (1969).
92) Peer, H. G., Schors, A.: Rec. Trav. Chim. 86, 161 (1967).
93) Petrov, A. A., Fedorova, A. V.: Usp. Khim. 33, 3 (1964).
94) Pilgrim, K., Korte, K.: Tetrahedron Letters 883 (1962).
95) Racanelli, P., Pantini, G., Immirzi, A., Allegra, G., Porri, L.: Chem. Commun. 361 (1969).
96) Ried, W., Käppeler, W.: Ann. Chem. 687, 183 (1965).
97) — Mengler, H.: Ann. Chem. 678, 95 (1964).
98) Roberts, J. D., Sharts, C. M.: Org. Reactions 12, 1 (1962).
99) Roedig, A.: Angew. Chem. Intern. Ed. Engl. 8, 150 (1969).
100) — Bischoff, F.: Naturwissenschaften 49, 448 (1962).
101) — — Heinrich, B., Märkl, G.: Ann. Chem. 670, 8 (1963).
102) — Detzer, N.: Ann. Chem. 710, 1 (1967).
103) — — Ann. Chem. 710, 7 (1967).
104) — — Angew. Chem. Intern. Ed. Engl. 7, 471 (1968).
105) — — Angew. Chem. Intern. Ed. Engl. 7, 472 (1968).
106) — — Friedrich, J. H.: Angew. Chem. Intern. Ed. Engl. 3, 382 (1964).
107) — Hagedorn, F.: Ann. Chem. 683, 30 (1965).
108) — Märkl, G., Heinrich, B.: Angew. Chem. Intern. Ed. Engl. 2, 47 (1963).
109) Roth, W. R., Friedrich, A.: Tetrahedron Letters 2607 (1969).
110) Salem, L.: J. Am. Chem. Soc. 90, 543, 553 (1968).
111) — Chem. Brit. 5, 449 (1969).

J. E. Baldwin and R. H. Fleming

112) Scharpen, L. H., Laurie, V. W.: J. Chem. Phys. *49*, 3041 (1968).
113) Seebach, D.: Fortschr. Chem. Forsch. *11*, 177 (1969).
114) Servis, K. L., Roberts, J. D.: J. Am. Chem. Soc. *87*, 1339 (1965).
115) Sethi, D. S., Joshi, G. C.: Can. J. Chem. *46*, 2632 (1968).
116) Sharma, R. K., Shoulders, B. A., Gardner, P. D.: J. Org. Chem. *32*, 241 (1967).
117) Shechter, H., Bleiholder, R. F.: J. Am. Chem. Soc. *90*, 2131 (1968).
118) Shingu, K., Hagishita, S., Nakagawa, M.: Tetrahedron Letters 4371 (1967).
119) Sisenwine, S. F., Day, A. R.: J. Org. Chem. *32*, 1770 (1967).
120) Skattebøl, L., Solomon, S.: J. Am. Chem. Soc. *87*, 4506 (1965).
121) Slobodin, Ya. M., Khitrov, A. P.: Zh. Obshch. Khim. *33*, 153 (1963).
122) — — Zh. Obshch. Khim. *33*, 2822 (1963).
123) — — Zh. Organ. Khim. *1*, 1531 (1965).
124) Staab, H. A., Kurmeier, H. A.: Chem. Ber. *101*, 2697 (1968).
125) Taylor, D R.: Chem. Rev. *67*, 317 (1967).
126) — Warburton, M. R.: Tetrahedron Letters 3277 (1967).
127) — Wright, D. B.: Chem. Commun. 434 (1968).
128) Tobey, S. W., West, R.: J. Am. Chem. Soc. *88*, 2478 (1966).
129) Untch, K. G., Martin, D. J.: J. Am. Chem. Soc. *87*, 4501 (1965).
130) van Remoortere, F. P., Boer, F. P.: Angew. Chem. Intern. Ed. Engl. *8*, 597 (1969).
131) Veniard, L., Benaim, J., Pourcelot, G.: Compt. Rend., Ser. C *266*, 1092 (1968).
132) Vrieze, K., Volger, H. C., Gronert, M., Praat, A. P.: J. Organometal. Chem. (Amsterdam) *16*, P 19 (1969).
133) Walbrick, J. M., Wilson, J. W., Jr., Jones, W. M.: J. Am. Chem. Soc. *90*, 2895 (1968).
134) Waters, W. L., Caserio, M. C.: Tetrahedron Letters 5233 (1968).
135) — Kiefer, E. F.: J. Am. Chem. Soc. *89*, 6261 (1967).
136) — Linn, W. S., Caserio, M. C.: J. Am. Chem. Soc. *90*, 6741 (1968).
137) Wedegaertner, D. K., Millam, M. J.: J. Org. Chem. *33*, 3943 (1968).
138) Weinstein, B., Fenselau, A. H.: Tetrahedron Letters 1463 (1963).
139) — — J. Chem. Soc. C 368 (1967).
140) — — J. Org. Chem. *32*, 2278, 2988 (1967).
141) Wilson, A., Goldhamer, D.: J. Chem. Educ. *40*, 504, 599 (1963).
142) Woodward, R. B., Hoffmann, R.: Angew. Chem. Intern. Ed. Engl. *8*, 781 (1969).
143) Wright, J. S., Salem, L.: Chem. Commun. 1370 (1969).

Received January 7, 1970

Nitrogen Inversion
Experiment and Theory

Prof. Dr. J. M. Lehn

Institut de Chimie, Université de Strasbourg, Strasbourg, France

Contents

1. Introduction

Tricoordinated atomic sites AXYZ (where X,Y,Z represent the atoms directly linked to A) in polyatomic molecules may have two types of local geometry: the A site may be pyramidal or planar. In AX_3 systems the corresponding local symmetry is respectively C_{3v} and D_{3h}. Bicoordinated sites AXY may be either bent or linear, corresponding to C_{2v} and $D_{\infty h}$ symmetries in the AX_2 case. If the stable form of AXYZ is pyramidal, there are two enantiomeric pyramidal forms (configurations) which may interconvert by passing through a planar transition state:

Similarly, two bent planar AXY forms may be interconverted through a linear transition state:

If the AXYZ and AXY sites are more stable respectively for a planar and linear structure, the molecular state is unique; this case will only be considered here as a special case where molecular structure imposes this geometry upon systems which "normally" are pyramidal or bent[a].

These interconversions occur by either *pyramidal* or *planar inversion* at the atomic site A, going over an energy barrier at the transition state. Two questions may then be asked:

1. *How* are the inversion processes, and especially the energy barriers hindering the processes, affected by the structure of the molecule containing the inverting site?

2. *Why* are there barriers to inversion (i. e. what makes the energy of the transition state higher) and why do specific molecular structural factors lead to the observed effects?

The answer to the first question is of an empirical nature and will be sought in terms of the nature of the inversion process, the structural factors affecting it and the rationalization of the results through more or less empirical or semi-empirical concepts and schemes.

---

[a] The influence of the nature of A on the preferred geometry of AXYZ and AXY compounds will not be considered here (see "Walsh's rules" [1]) Stable pyramidal forms may be expected for $A = C^-$, N, O^+, Si^-, P, S^+. . . . .

The answer to the second question is theoretical, as a knowledge of the detailed microphysical interactions between electrons and nuclei is a prerequisite to giving a unified *physical picture* of the inversion process and of the factors affecting it.

We shall limit ourselves to the case where the A site is a nitrogen atom, i.e. to *nitrogen inversion*. It is, however, clear that a number of conclusions may hold as well for $A = P$, C^-, O^+ etc. . . We shall consider in turn the nature of the nitrogen inversion process, the structural effects on barriers to inversion and the theoretical studies of the process.

We shall not discuss exhaustively all the numerous results published in recent years on the subject but merely consider those which are especially relevant to or have strong bearing on the present issues.

2. The Nitrogen Inversion Process

2.1. Types of Inversion Processes

As pointed out above, two types of processes may be considered:

a) *Pyramidal nitrogen inversion* of NXYZ compounds through a planar transition state, along the normal mode of bending vibration ν_0 (parallel to the axis of the pyramid when $X=Y=Z$):

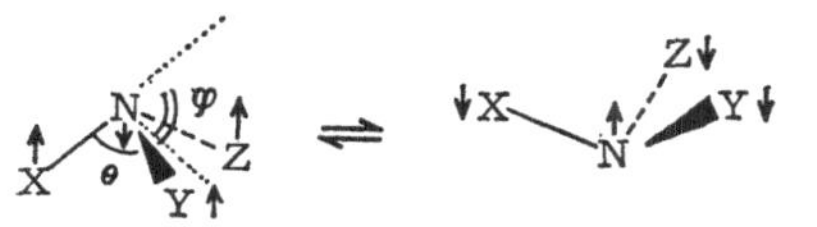

$$\tag{1}$$

More or less pyramidal nitrogen sites are found in amines and their derivatives (amides, haloamines, hydroxylamines . . .).

b) *Planar nitrogen inversion* of bent NXY compounds by in-plane wagging through a linear transition state:

$$\tag{2}$$

Such processes may be operative in imines and their derivatives (oximes...) azo compounds, diimides etc.

In a number of cases, these two types of interconversion may be brought about as well by rotation about bonds, e.g. rotation about the (X,Y,Z)-N single bonds or the $X=N$ double bond. This situation holds, for instance, in acyclic amines (methylamine, hydroxylamine . . .) and in imines. Furthermore, nitrogen inversion may in some cases occur in the same molecule together with ring inversion, for instance, in piperidine. It will thus sometimes be necessary to distinguish between *internal rotation* or *ring inversion* and *nitrogen inversion*. The ways of separating these processes will be discussed below. However, separation may not always be possible (either because the two processes are hindered by barriers of similar height or because there is only a single transition state) and the nature of the process may remain doubtful.

We shall centre the discussion on pyramidal nitrogen inversion, which has been studied more extensively, and treat planar nitrogen inversion separately.

2.2. Mechanisms and Rates of Inversion

The potential energy curve for the nitrogen inversion process in NX_3 is a symmetrical double minimum curve with an energy barrier V_{max} (Fig. 1).

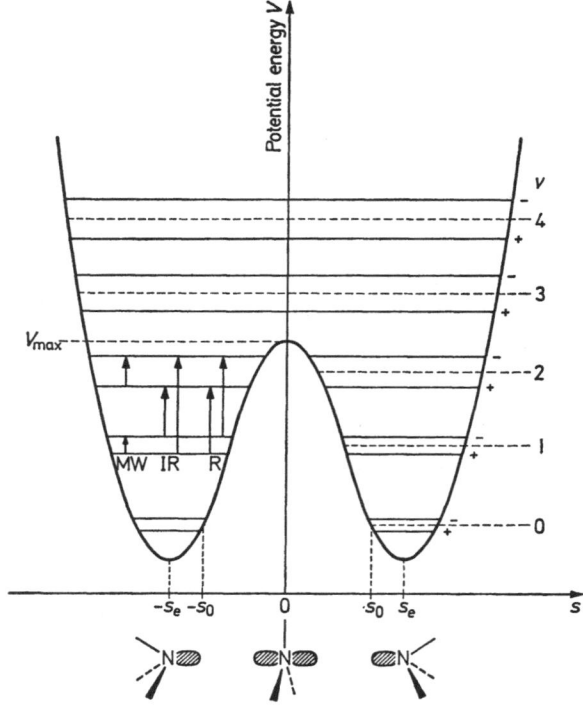

Fig. 1. Potential energy curve (as a function of the distance s from the N atom to the X_3 plane) and vibrational energy levels for the parallel vibrational mode v_0 of a pyramidal NX_3 molecule. The splittings into symmetric (+) and antisymmetric (−) levels are not drawn to scale. MW, IR and R represent types of transitions measured respectively by MicroWave, InfraRed and Raman spectroscopy. s_e is the equilibrium height of the NX_3 pyramid

Inversion may occur either classically by passing *over* the barrier or by quantum-mechanical tunneling *through* the barrier.

The rates for going over and through the barrier do not depend in the same way on its height and shape. Thus, as barrier heights are determined from inversion rates (see below), it is necessary to know which mechanism is operative or predominant in a particular case.

The *classical inversion mechanism* is a thermally activated process [2], activation energies being determined from the variation of inversion rates with temperature. The corresponding rates for passage over the barrier may be calculated from the absolute reaction rate theory [2]. The rate constant is given by the Eyring rate equation:

$$k = \varkappa \frac{k_B T}{h} \frac{F_{\neq}}{F_i} e^{-E_0/RT} \tag{3}$$

where $\varkappa$ is the transmission coefficient, k_B and h are respectively the Boltzmann and Planck constants, E_0 is the zero-point activation energy (see below and Fig. 2), $F_{\neq}$ and F_i are the partition functions of the *transition state* (TS) and initial state (*ground state* GS)[a] respectively.

When tunneling is occurring together with barrier crossing, the observed total rates are faster than the classical ones and thus the apparent activation energies are lower than the real barrier hindering inversion.

Inversion by Tunneling [3-10]

Quantum mechanically there is a finite probability that inversion may occur even when the vibrational energy of the molecule is lower than the potential barrier V_{max}. The vibrational wave functions for the parallel vibration [Eq. (1)] in the left (ψ_L) and right (ψ_R) potential minima penetrate the barrier and overlap to some extent. A given vibrational state is then described by a linear combination of ψ_L and ψ_R into a symmetrical ψ^+ and an antisymmetrical ψ^- function:

$$\psi^+ = \frac{1}{\sqrt{2}} (\psi_L + \psi_R)$$

$$\psi^- = \frac{1}{\sqrt{2}} (\psi_L - \psi_R) \tag{4}$$

Each vibrational level will be split into a symmetrical and an antisymmetrical level. The splitting will be the more pronounced the larger

---

[a] The term Ground State (GS) will be used for the more stable pyramidal or bent configuration in the electronic ground state of the system.

the interaction of the wave functions on both sides of the barrier; thus excited vibrational states show larger plittings than the ground state, the barrier being thinner, and its top lower above the energy of the level (see Fig. 1).

The *frequency of tunneling* ν_t through the barrier is related to the overlap of ψ_L and ψ_R and is proportional to the inversion splitting (energy difference) $\Delta E_i = E_i^- - E_i^+$ between the ψ^+ and ψ^- components of a given vibrational state:

$$\nu_t = 2 \, \Delta E/h \tag{5}$$

It may be shown that the rate of inversion by tunneling through a potential barrier decreases exponentially with the reduced mass μ of the system, the thickness s_i and the height V_{max} (i.e. the area) of the barrier 3–10):

$$\nu_t = \frac{2\nu_0}{\pi} A_i \qquad A_i = \exp\left\{-\frac{4\pi}{h} \int_0^{s_i} [2\,\mu\,(V(s) - \varepsilon_i)]^{1/2} ds\right\} \tag{6}$$

with ν_0: vibration frequency on each side of the barrier, ε_i, s_i: vibrational energy $(= (v + 1/2)h\nu_0$; v: vibrational quantum number) and height of the pyramid in the vibrational state i.

Assuming a Boltzmann distribution, the tunneling frequency (rate constant k_t of the unimolecular process) is:

$$\nu_t = k_t = \frac{2\nu_0}{\pi} \sum_i A_i \, \exp\,(-\varepsilon_i/kT) \, / \, \sum_i \exp\,(-\varepsilon_i/kT) \tag{7}$$

Table 1 lists some inversion splittings measured by *microwave spectroscopy*. The NH_3 molecule undergoes inversion by tunneling at a frequency of 4.10^{10} sec^{-1} and 2.10^{12} sec^{-1} in its ground and first excited vibrational states respectively. As the inversion barrier is about 6 kcal/mole (Table 1) the thermal rate at 300 °K would be only ca. 2.10^8 sec^{-1}. It is also seen that increasing the reduced mass by a factor of 2 as in ND_3 decreases the inversion rate by a factor of 11. Molecules with small and thin barriers (slightly pyramidal molecules) like H_2N-CN, H_2N-CHO undergo very fast tunneling (Table 1).

The same is true for most primary and secondary amines (except those having high barriers like aziridine, oxaziridine . . .).

The tunneling frequency is expected to be much smaller when the inversion barrier becomes high and thick and when the reduced mass of the systems increases. In such cases (for instance, tertiary amines, aziridines . . .) the rate of tunneling is expected to add little to the rate

Table 1. *Inversion splittings ΔE (in MHz) and inversion barriers V_{max} (in kcal/mole) in ammonia and derivatives*

Compound No.	ΔE (method[a]); references)[b]		Inversion Barrier V_{max} (method[a]); references)
	Vibrational Ground State $(v = 0)$	First excited vibrational State $(v = 1)$	
1 NH_3	23,786 (MW; [3])	1,095,000	5.77 (MW; [13])
2 NH_2D	12,182 (MW; [3])	592,000	—
3 NHD_2	5,111 (MW; [3])	295,000	—
4 ND_3	1600 (MW; [3])	117,000	—
5 CH_3NH_2	28,605 (IR; [30])	948,000	4.8 (IR; [30])
6 CD_3NH_2	24,000 (IR; [30])	540,000	—
7 CH_3ND_2	2310 (IR; [30])	390,000	—
8 CD_3ND_2	2220 (IR; [30])	—	—
9 $(CH_3)_2NH$	2646 (MW; [31])	—	4.4 ± 1.1 (MW; [31])
10 $H_2N—C_6H_5$	<3,000,000 (MW; [32])	—	~2 (estimated from non-planarity [32])
11 $H_2N—CN$	5,340,000 (IR; [33]) 1,500,000 (MW; [34])	— —	1.9 (IR; liquid state; [33]) 2.03 (MW; [34])
12 $D_2N—CN$	2,610,000 (IR; [33])	—	—
13 $H_2N—CHO$	—	—	1.1 (MW; [35])
14 $H_2N—NO_2$	—	—	2.7 (MW; [32])
15 NHF_2	<0.15 (MW; [36])	—	—
16 Aziridine	<0.025 (MW; [12b]) <0.015 (MW; [12a])	— —	>12 (MW; [12b]) >11.6 (MW; [12a])

[a]) MW:Microwave spectroscopy; IR:Infrared spectroscopy.
[b]) Tunneling rates are given by $\nu_t = 2\ \Delta E$.

for going over the barrier (thermal inversion rate) [11,12] and its effect is essentially to lead to a somewhat lower apparent thermal barrier by rounding off the top of the real barrier.

Furthermore it may be seen from Eq. (3) and (7) that the classical rates k are more strongly temperature dependent than the tunneling rates k_t. Thus the ratio k_t/k *decreases rapidly as the temperature increases* and the tunneling corrections generally become small at room temperature and above.

In cases where the double minimum inversion potential is unsymmetrical it has been shown that the tunnel frequencies are strongly decreased even for slight dissymetries of the potential curve [8-10].

Inversion of planar nitrogen sites also occurs in a double minimum potential and thus the above discussion applies as well. However, barriers to inversion are generally high (>20 kcal/mole; see below) and the effect of tunneling should be negligible (except for certain carbodiimides; see below).

2.3. Determination of Barriers to Nitrogen Inversion

Two types of methods for determining the height of barriers to nitrogen inversion may be distinguished:

A. the methods measuring *tunneling rates* by observing their spectroscopic effects:

B. the methods measuring *total inversion rates*, which are the classical rates when tunneling is negligible.

A. Inversion Barriers from Rates of Tunneling

As discussed above, the *vibrational energy levels* in a double minimum potential curve are split into *symmetric* and *antisymmetric states*. Spectroscopic transitions between these levels may occur.

Fig. 1 indicates which transitions may be observed in microwave, infrared and Raman spectroscopy. The line frequencies and doublet separations are related to the tunneling frequency.

In order to calculate the height V_{max} of the potential barrier from the measured tunneling frequencies [see Eq. (6), (7); ref. [3-14]] it is necessary to know the *geometry of the molecule* (especially the height of the nitrogen pyramid), which may also be obtained from analysis of the microwave spectra. Fitting then a barrier shape function [13] to the observed spectral data leads to the value of V_{max}. Thus an approximate shape of the potential curve is obtained.

It is clear that the tunneling rates have to be high enough for spectroscopic transitions or doublings to be experimentally observable. Thus these methods are limited to relatively low inversion barriers (below ca. 6 kcal/mole); for instance, the inversion barrier in aziridine cannot be determined in this way and only an upper limit (30 kHz) to the tunneling rate is obtained giving a lower limit for the barrier [12].

B. Inversion Barriers from Total Inversion Rates

Nitrogen inversion rates have been measured by essentially three types of methods:

a) Spectroscopic methods: essentially Nuclear Magnetic Resonance (NMR);

b) Kinetic methods using epimerization or racemization processes;

c) Relaxation methods: some dielectric relaxation data have been interpreted in terms of nitrogen inversion processes.

a) NMR Determination of Rate Constants [14–16]

The NMR method (Dynamic NMR: DNMR) [16] of determining rate constants covers a range of rates corresponding to barrier heights from 5–6 kcal/mole to 20–25 kcal/mole. This range lies in between the domains where microwave spectroscopy (below) or classical kinetic methods (above) may be used.

When nitrogen inversion is occurring, temperature-dependent NMR spectra may be observed and the inversion rates may be extracted by analysing the spectral changes. Recent reviews cover the use of NMR for studying rate processes and we refer to them for more details about the methods, their limitations and the sources of errors [14–16]. With the presently available complete lineshape analysis techniques together with spectrum simplification procedures (by double irradiation, deuteration . . .) accurate determination of rate constants is possible over the temperature range where the spectral changes are appreciable (especially above and below the coalescence temperature).

Errors become more important in both limiting ranges of slow or fast (on the NMR time scale) rates. *Spin-echo* [14–16] or *multiple resonance* [17a] *methods* may be used to extend the range and thus to increase the accuracy of the results.

b) Kinetic Methods

When the inversion barriers are above ca. 23 kcal/mole, the isolation of *epimers* (invertomers) or *optically active substances* becomes feasible. The first-order rates for the unimolecular epimerization or racemization processes may then be measured at several temperatures by following the interconversion processes spectroscopically (NMR, UV, IR) or polarimetrically. The treatment of the data is straightforward [18].

c) Relaxation Methods

Dielectric relaxation measurements on some amines (aniline, N,N-dimethylaniline, benzidine . . .) display two relaxation times: a longer one (of the order of $20–30.10^{-12}$ sec) attributed to overall molecular reorientation, and a shorter one (of the order of 1.10^{-12} sec) which may be interpreted as arising from an intramolecular process and has been attributed to nitrogen inversion [19–22], although this attribution is not unequivocal (rotation about the N—C bond may also contribute to the observed relaxation).

d) Calculation of Inversion Barriers

The *free energy of activation*, $\Delta G^{\neq}$, at temperature T may be calculated from the inversion rate at temperature T using the Eyring rate equation [2]:

$$k = \varkappa \frac{k_B T}{h} \exp\left(-\Delta G^{\neq}/RT\right) \tag{8}$$

where the transmission coefficient $\varkappa$ is generally taken as unity.

With $\Delta G^{\neq} = \Delta H^{\neq} - T \Delta S^{\neq}$ ($\Delta H^{\neq}$: enthalpy and $\Delta S^{\neq}$: entropy of activation), $\varkappa = 1$ and using the numerical values of the constants and decimal logarithms, (8) becomes:

$$\Delta G^{\neq} = 4.57 \, T \, [10.32 - \log(k/T)]$$

$$\log (k/T) = 10.32 - \frac{\Delta H^{\neq}}{4.57 \, T} + \frac{\Delta S^{\neq}}{4.57} \tag{9}$$

If k has been determined at different temperatures one may then plot $\log(k/T)$ versus $1/T$, giving a straight line with slope $-\Delta H^{\neq}/4.57$ and with intercept $(10.32 + \Delta S^{\neq}/4.57)$. Thus the enthalpy and entropy of activation are obtained. One can also use the Arrhenius equation and plot $\log k$ versus $1/T$:

$$\log k = -E_a/4.57 \, T + \log A \tag{10}$$

The slope of the straight line obtained gives the activation energy E_a and its intercept gives the frequence factor A.

It is now of importance to know the relations between the Eyring and the Arrhenius activation parameters and the potential barrier V_{max}. One has [2]:

$$\Delta H^{\neq} = E_a - RT \tag{11}$$

and

$$\Delta S^{\neq} = R \left[\ln\left(\frac{h A}{k_B T}\right) - 1\right] \tag{12}$$

$$= 4.57 \log (A/T) - 49.20 \tag{13}$$

One may show that the relation between the zero-point activation energy E_0 (3) and the activation enthalpy $\Delta H^{\neq}$ is [2]:

$$\Delta H^{\neq} = E_0 + RT^2 \frac{d \ln(F^{\neq}/F_i)}{dT} \tag{14}$$

and

$$E_0 = V_{max} + N \sum \tfrac{1}{2} h\nu^{\neq} - N \sum \tfrac{1}{2} h\nu_i \tag{15}$$

where $\nu^{\neq}$ and ν_i are the frequencies of the vibrations of the transition and initial states respectively and N is the Avogadro number (see also Fig. 2).

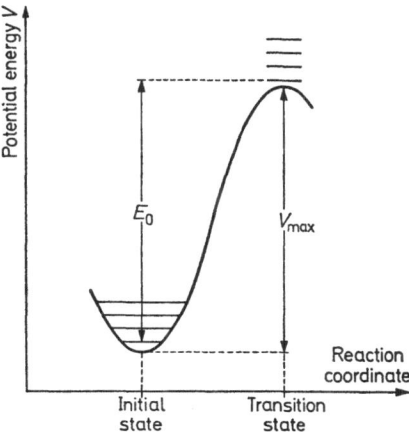

Fig. 2. Relation between the zero-point activation energy E_0 and the classical potential barrier V_{max}

As the variation of the logarithm of the partition function ratio with temperature is probably very small [b], one may write:

$$\Delta H^{\neq} = E_0 \tag{16}$$

Now, it seems reasonable that, although $N \sum \frac{1}{2} h\nu$ changes from initial state to transition state, its variation is probably small compared to the barrier height. Then:

$$\Delta H^{\neq} \sim V_{max} \tag{17}$$

and $\Delta H^{\neq}$ may be considered as a good approximation to the potential barrier to inversion V_{max}. As E_a and A are not temperature-independent parameters, it seems preferable to assume temperature-independent

---

[b] For a unimolecular reaction like nitrogen inversion the translational, rotational and vibrational partition functions per degree of freedom may be assumed not to differ greatly in the initial and transition states. Then $F_{\neq}/F_i$ may be reduced to $1/f_v$ where f_v is the partition function for one vibrational degree of freedom. $\ln (1/f_v)$ is of the order of -1 to 0 and should not change much with temperature 2.18).

$\Delta H^{\neq}$ and $\Delta S^{\neq}$ values. In the following sections only $\Delta G^{\neq}$, $\Delta H^{\neq}$ and $\Delta S^{\neq}$ values will be used and results from the literature will be converted into these parameters.

e) Limitations

There are several sources of errors which limit the accuracy of barrier determination (in addition to the approximation discussed above). Kinetic measurements deal with high barriers to inversion so that tunneling effects are certainly negligible and reliable $\Delta H^{\neq}$ values may be obtained from careful studies.

Most recent results about the activation energies for nitrogen inversion have been obtained by NMR studies. Due to the errors in rate measurement (see above and ref. [14-16,23]), the $\Delta H^{\neq}$ and $\Delta S^{\neq}$ values obtained from Eyring plots bear large errors and are quite unreliable.

At the present time no really high-quality, NMR-determined, $\Delta H^{\neq}$ and $\Delta S^{\neq}$ values for nitrogen inversion are available in the literature[c]. $\Delta S^{\neq}$ values from -20 ro $+30$ eu have been reported! In addition many studies just give $\Delta G^{\neq}$ values and often only $\Delta G_c^{\neq}$ for the coalescence temperature (where however accuracy is highest). As in the following sections we try (and need to try!) to introduce somewhat more coherence between results given in the literature, we shall proceed as follows:

1. inversion barriers will be reported as $\Delta H^{\neq}$ values[d];

2. in all cases, these $\Delta H^{\neq}$ values will be calculated from the results in the literature (preferably $\Delta G_c^{\neq}$ values which are probably less inaccurate than other data [23]) using a $\Delta S^{\neq}$ value of $+5$ eu (even when larger $\Delta S^{\neq}$ values are reported). Large $\Delta S^{\neq}$ values are not expected for intramolecular processes and arise probably from systematic errors[e]. Small positive $\Delta S^{\neq}$ values have been obtained in recent kinetic studies of inversion barriers (see below)[f]. We think that in the present state of affairs, the procedure we adopt is more reliable (or less unreliable!) than taking and comparing the reported results as they are.

---

[c] This is clearly an important gap. By high quality we mean of a quality comparable to the Anet and Bourn [17a] study of ring inversion in cyclohexane.

[d] One could also report the barriers as $\Delta G^{\neq}$ values at the *same* temperature (using again $\Delta S^{\neq} = +5$ eu); both procedures are presumably equivalent with respect to accuracy but $\Delta H^{\neq}$ values are more directly related to barrier heights (see above).

[e] Of course one cannot just dismiss large $\Delta S^{\neq}$ values which might be present in special cases. However, with the wealth of merely *qualitative* data, we feel that the procedure we adopt is probably the less unreliable one! Recent accurate determinations of barriers to internal rotation by NMR spectroscopy yield small or nearly zero activation entropies [17b].

[f] The $\Delta H^{\neq}$ values obtained for $\Delta S^{\neq} = 0$ and $\Delta S^{\neq} = +5$ eu would differ by "only" ca. 10%.

As a further source of errors, tunneling may not be negligible for inversion barriers in the low range (*ca.* 5—6 kcal/mole) of the NMR method. Tunneling increases the total rate over the classical rate, and this effect increases as the temperature decreases. Thus an Eyring (or Arrhenius) plot should curve at low temperature [24] (Fig. 3). As a consequence, "straight" lines drawn through the experimental points will have a smaller slope and a lower intercept, thus leading to smaller apparent $\Delta H^{\neq}$ (or E_a) and lower apparent $\Delta S^{\neq}$ (or log A) values. The contribution of tunneling is expected to be relatively unimportant in most compounds because of the heavy reduced masses and of the relatively high barriers (> 8 kcal/mole) involved. In addition, the error introduced by the tunneling curvature of Eyring plots just adds to the errors due to the method, which are at a maximum at the low (and high) temperature edge of the plot. This factor again favours the use of $\Delta G^{\neq}$ values and of a fixed $\Delta S^{\neq}$ value until more accurate experimental data become available [g].

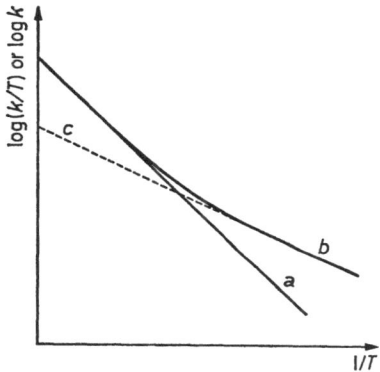

Fig. 3. Temperature dependence of reaction rates (Eyring or Arrhenius plots): *a* classical rates; *b* total rates taking into account the effect of tunneling; *c* type of extrapolated curve obtained from rate measurements over a limited temperature range

The inversion barriers determined by dielectric relaxation are subject to all the above sources of errors, and tunneling certainly becomes of importance as the barriers measured are very low (of the order of a few

---

[g] In fact the $\Delta S^{\neq}$ values also incorporate the possible variations of the transmission coefficient $\varkappa$, for which a fixed value of 1 is used. Thus in addition to the very large errors involved in determining $\Delta S^{\neq}$, the physical meaning of Eyring's entropy of activation is not clear [25].

kcal/mole). In addition, the smallness of the barriers introduces another inaccuracy factor in the determination of Arrhenius activation energies. When the energy barrier becomes comparable to RT, the apparent activation energy E_a obtained from Arrhenius plots cannot be taken as a measure of the energy barrier; at the limit no conclusions can be drawn about the value or the existence of a barrier [26,27].

In conclusion, the following sections will make use nearly exclusively of inversion barriers determined by microwave spectroscopy and activation enthalpies obtained from DNMR and classical kinetic studies.

As medium effects may be important (see below), attention will be paid to selecting the literature results (when available) which minimize such effects, in order to deal with as comparable values as possible [h].

3. Structural Effects on Nitrogen Inversion Barriers

Barriers to nitrogen inversion are strongly dependent on molecular structure and, as such, they may be considered as probes for understanding structural effects in molecules.

In the present section, structural factors will be described in terms of empirically or semi-empirically defined concepts using the language of molecular mechanics (*steric effects*) or of simple electronic structure theory (*electronic effects*). The separation of these various effects is often not possible and they may overlap in the same molecule [93]. Thus, an analysis in terms of structural factors may seem artificial and in the present section these factors will mainly be used for correlating experimental data, with the hope that such correlations may present a high enough degree of confidence to be usable in estimating before-hand inversion rates or barriers. Tables 1—6 (see pp. 317, 326—341, 352—355) list selected experimental results obtained for nitrogen inversion in a variety of molecules [a].

3.1. Steric Effects

Essentially two types of steric effects [93] will be considered:
A) nonbonded interactions;
B) ring strain due to the inclusion of the nitrogen site in a cyclic system.

---

[h] Rates of nitrogen inversion determined by the method of Saunders and Yamada [28] are for water solutions, and thus incorporate an appreciable medium effect (see also Section 4).

[a] For earlier reviews on nitrogen inversion see: [29], general review; [37], nitrogen inversion in aziridines.

In looking for such effects, only compounds where the nitrogen substituents are expected to have similar electronic properties will be considered. During the inversion process the XNY angle θ [Eq. (1)] increases towards 120°; thus, structural factors facilitating or hindering this angle opening will affect the barrier to inversion in an opposite fashion.

A. Non-Bonded Interactions

Van der Waals potential functions for non-bonded interactions display an attractive and a repulsive region [93]. Attractive interactions are small, too small to lead to a detectable effect on nitrogen inversion barriers in the present state of data accuracy. The repulsive portion of the curve is however very steep. Thus the *presence of bulky substituents* leads to appreciable nonbonded repulsions which are stronger in the pyramidal than in the planar state, where repulsions are partially relieved by the opening of the angle θ. As a consequence the pyramidal state is destabilized with respect to the planar TS and the inversion barrier is expected to decrease.

Thus in aziridines *(17)* and *(22)* replacing a $N-CH_3$ by a $N-C(CH_3)_3$ group leads to a ca. 2.5 kcal/mole barrier decrease. The same structural change produces an even larger effect in the oxaziridines *(80)* and *(81)* where the barrier decrease is larger than 6 kcal/mole.

The introduction of a gem-dimethyl group α to the inverting nitrogen site leads to a ca. 1.0—1.5 kcal/mole barrier decrease in aziridine derivatives (see *19, 35; 21, 34; 57, 63*); a β gem-dimethyl group lowers the barrier by 0.5—1.0 kcal/mole in pyrrolidine derivatives (see *93, 95; 94, 99*).

Various other examples may be found in the Tables. In particular, when the two invertomers are of different energy, the smallest inversion barrier corresponds to the epimer having the bulkiest groups *cis* to one another (see *74, 75, 78, 79, 83, 84*) and in such cases often only one invertomer may be observed in the NMR spectrum as, for instance, in N-ethyl-*cis*-2,3-dimethylaziridine [40]. The 3 kcal/mole barrier lowering found in the N-chloroaziridine *(75)* as compared to *(74)* arises from the more important steric repulsions present in *(75)*.

B. Ring Strain

Internal strain in cyclic systems [93, 95] incorporates angle strain (Baeyer strain), torsional strain (Pitzer strain) and transannular strain. In the case of nitrogen inversion, large effects are expected to arise from *angle strain*[b], the nature of the cyclic system imposing a value of the angle θ

---

[b] It should be noted that as an alternative to the molecular mechanics picture of angle strain, a hybridization picture may also be given (see [96] and below).

Table 2. *Activation parameters for pyramidal nitrogen inversion from NMR data: Aziridines and substituted aziridines*[a]

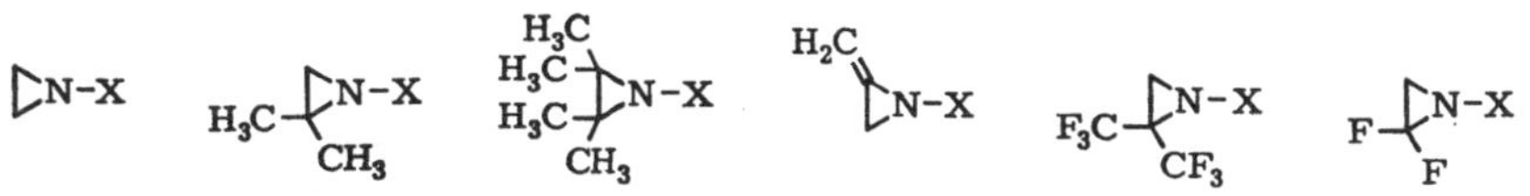

Structural types:

$\underset{A}{\triangle\!N\text{-}X}$ $\underset{B}{H_3C\text{-}\triangle\!N\text{-}X\ (CH_3)}$ $\underset{C}{\substack{H_3C\\H_3C\\H_3C}\triangle\!N\text{-}X\ (CH_3)}$ $\underset{D}{H_2C\!=\!\triangle\!N\text{-}X}$ $\underset{E}{F_3C\ \triangle\!N\text{-}X\ (CF_3)}$ $\underset{53}{F,F\ \triangle\!N\text{-}X\ (F)}$

Compound №	Structural Type	Substituent X on Nitrogen	Solvent[b]	Ref.	Literature data[c]			$\Delta G_c^{\neq}$ kcal/mole	$\Delta H^{\neq e}$ ($\Delta S^{\neq} = +5$ eu) kcal/mole
					T_c (°C)	k_c (sec^{-1})	Other data		
17	A	—CH₃	neat	39)	—	—	$E_a = 19\ \log A = 11$ ($\Delta H^{\neq} = 18.2$; $\Delta S^{\neq} = -10.7$)	$\Delta G_{100} = 22.3(?)$	24.2(?)
			TCE	46)	115°	98	—	19.4	21.3
18	A	—C₂H₅	neat	40)	108°	60	—	19.4	21.3
			D₂O	40)	>145°	<60	—	>21.3	>23.4
19	A	—CH₂C₆H₅	neat	40)	105°	60	—	19.2	21.1
20	A	—CH₂CH₂C₆H₅	neat	40)	96°	69	—	18.6	20.4
21	A	—C₆H₁₁	neat	40)	95°	51	—	18.8	20.6
22	A	—C(CH₃)₃	neat	37,41)	52°	24	—	17.0	18.6
23	A	—Adamantyl	CCl₄	37,41)	41°	10	—	17.0	18.6
24	A	—CH₂OMe	CCl₄	42)	30°	16.5	—	16.0	17.5

25	A	−CH₂N(CH₃)₂	CCl₄	42)	65°	28	—	17.6	19.3
26	A	−CH₂OH	CCl₄	42)	83°	21.5	—	18.8	20.6
27	A	−C₆H₅	CS₂	43)	− 40°	50	—	11.7	12.9
28	A	−COOCH₃	VC	43)	−138°	~10	—	7.1	7.8
29	A	−CON(CH₃)₂	VC	43)	− 86°	~10	—	9.9	10.8
30	A	−COCH₃	VC	43)	<−160°	—	—	< 6	—
31	A	−COC₆H₅	—	44)	<−155°	—	—	<6	—
32	B	−CH₃	CDCl₃	45)	CH₃: 67.5°	7	—	18.5	20.2
			neat	45)	CH₃: 61°	19	—	17.6	19.3
33	B	−CH(CH₃)₂	CDCl₃	46)	CH₃: 54°	12	—	17.6	19.2
					CH₂: 75°	75	—	17.5	19.2
34	B	−C₆H₁₁	CDCl₃	46)	CH₃: 51°	10.5	—	17.5	19.1
					CH₂: 73°	72	—	17.4	19.1
35	B	−CH₂−C₆H₅	CDCl₃	46)	CH₂: 79°	73	—	17.7	19.5
36	B	−C₆H₄−pOCH₃	CF₂Cl₂	47)	CH₃: −26°	50	ΔH≠ = 15.2 ± 1; ΔS≠ = 11.4 ± 6	12.5	13.7
			CDCl₃	46)	CH₃: −25°	59	—	12.4	13.6
					CH₂: −27°	46	—	12.4	13.6
37	B	−C₆H₅	CF₂Cl₂	47)	CH₃: −49°	56	ΔH≠ = 13.6 ± 1; ΔS≠ = 10.4 ± 6	11.2	12.3
38	B	−C₆H₄−pCl	CF₂Cl₂	47)	CH₃: −53°	56	—	11.0	12.1
39	B	−C₆H₄−pCF₃	CF₂Cl₂	47)	CH₃: −72°	60	—	10.0	11.0
40	B	−C₆H₄−pNO₂	CHFCl₂	47)	CH₃: −107°	49	—	8.2	9.0
41	B	−C₆H₄−mCF₃	CF₂Cl₂	47)	CH₃: −59°	55	—	10.7	11.8
42	B	−COOMe	CHFCl₂	46)	<−140°	—	—	< 7	—

Table 2 (continued)

Compound No	Structural Type	Substituent X on Nitrogen	Solvent[b]	Ref.	Literature data[c] T_c (°C)	k_c (sec^{-1})	Other data	$\Delta G_c^{\neq c}$ kcal/mole	$\Delta H^{\neq c}$ ($\Delta S^{\neq} = +5$ eu) kcal/mole
43	C	–H	CCl$_4$	48)	52°	25.5	$E_a = 11-11.9$	17.0	18.6
44	C	–D	CCl$_4$	48)	68°	25.5	$E_a = 14.3-15.0$	17.9	19.6
45	D	–CH$_3$	neat	38)	–	–	$E_a = 6.4; \log A = 9$	10.1 (assuming $T_c \sim -50°$)	1.12
46	D	–C$_2$H$_5$	neat	40)	–65°	67	–	10.3	11.3
			0.01 NaOH in CH$_3$OH	40)	–25°	67	–	12.3	13.5
47	A	–CF$_2$–CFHCF$_3$	CCl$_4$	50)	11°	–	$E_a = 9.1$; $\log A = 9.6$ ($\Delta H^{\neq} = 8.5$; $\Delta S^{\neq} = -16.5$)	13.2	14.6
48	A	–CF$_2$–CH(CF$_3$)$_2$	CCl$_4$	50)	–13°	–	$E_a = 6.9$; $\log A = 8.3$ ($\Delta H^{\neq} = 6.4$; $\Delta S^{\neq} = -22.3$)	12.2	13.5
49	B	–CF$_2$–CFHCF$_3$	CCl$_4$	50)	9°	–	$E_a = 6.8$; $\log A = 7.8$ ($\Delta H^{\neq} = 6.2$; $\Delta S^{\neq} = -24.5$)	13.2	14.6

50	*B*	$-CF_2-CH(CF_3)_2$	CCl_4	50)	CH_3: −39°	—	$E_a = 5.8$; log $A = 7.8$ ($\Delta H^{\neq} = 5.3$; $\Delta S^{\neq} = -24.6$)	11.3	12.5
51	*E*	$-CH_3$	NB	51)	40°	2200	$E_a = 7.0$; log $A = 6.1$ ($\Delta H^{\neq} = 6.4$; $\Delta S^{\neq} = -32.7$)	13.5 (from k_c) (?) 16.6 (from E_a) (?)	15.1(?) 18.2(?)
52	*E*	$-C_6H_5$	CCl_4	51)	<− 40°	—	—	—	—
53	—	$-CF_3$	neat	52)	25°	—	$E_a = 5.5$; log $A = 8.7$ ($\Delta H^{\neq} = 5.0$; $\Delta S^{\neq} = -20.4$	10.1	11.3

a) See also ref. 37 for a list of inversion rates in aziridines.

b) VC: vinylchloride; NB: nitrobenzene; HCB: hexachlorobutadiene; DCB: dichlorobenzene; Ac: acetone; Py: pyridine; CP: cyclopropane; TCE: tetrachloroethylene

c) k_c: inversion rate (in sec^{-1}) at coalescence temperature T_c (degrees Celsius); A (sec^{-1}), E_a (kcal/mole) preexponential factor and Arrhenius activation energy; $\Delta H^{\neq}$ (kcal/mole), $\Delta S^{\neq}$ (e.u.): Eyring enthalpy and entropy of activation given in the literature or calculated from E_a and A. The nature of the signals for which the data are reported is given as CH_3: $k_c = \ldots$, $T_c \ldots$. k_c is calculated from $k_c = \pi\, \Delta\nu/\sqrt{2}$ (coalescence of two singlets) or $k_c = \pi\, (\Delta\nu^2 + 6\, J^2)^{1/2}/\sqrt{2}$ (coalescence of an AB pattern), where $\Delta\nu$ is the chemical shift difference in Hz and J the coupling constant when exchange is slow; in some cases the temperature dependence of $\Delta\nu$ has been taken into account. The accuracy of k_c and T_c is not the same in all cases, but it is reasonable to expect that in most cases ΔG_c should be accurate to better than $\pm$ 0.5 kcal/mole (generally $\pm$ 0.3 kcal/mole). Questionable (because of lack of literature data, inaccurate or out of date measurements ...) values are noted: (?).
It should be noted that the absence of line splittings at low temperature may also be due to too small chemical shift differences and does not necessarily imply fast exchange.

Table 3. *Activation parameters for pyramidal nitrogen inversion from NMR data: Aziridines and derivatives containing heteroatoms linked to the nitrogen atom*[a])

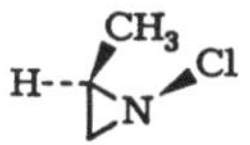

74

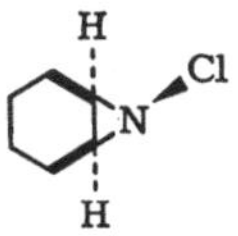

75

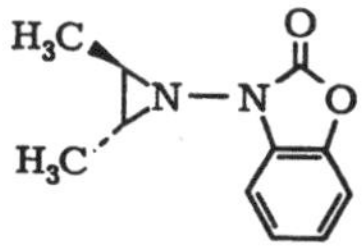

76

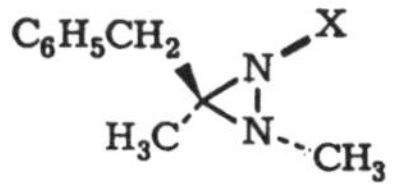

77 X = CH₂C₆H₅
78 X = H

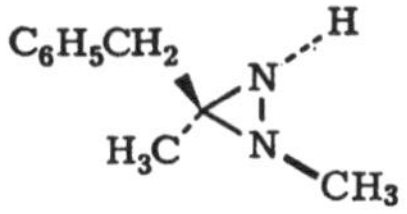

79

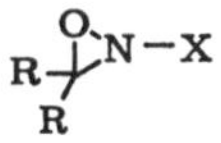

80 R = C₆H₅; X = CH₃
81 R = C₆H₅; X = C(CH₃)₃
82 R = H; X = C(CH₃)₃

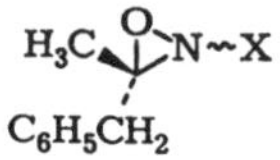

83 X = ⎯⎯ CH₃
84 X = --- CH₃

Compound No	Structural Type	Substituent X on Nitrogen	Solvent[a]	Ref.	T_c (°C)	k_c (sec^{-1})	Other data	$\Delta G_c^{\neq[a]}$ kcal/mole	$\Delta H^{\neq[a]}$ ($\Delta S^{\neq} = +5$ eu) kcal/mole
					Literature data[a]				
54	A	$-NH_2$	p-Xylene	53)	>150°	—	—	>22	—
55	A	$-SO_2CH_3$	$CDCl_3$	43)	−25°	25	—	12.8	14.0
56	A	$-SOC_6H_5$	$CDCl_3$	54)	0°	79	—	13.5	14.9
57	A	$-SC_6H_5$	$CDCl_3$	54)	−11°	71	—	13.0	14.3
58	A	$-P(O)(C_6H_5)_2$	CH_2Cl_2	54)	−108°	48	—	8.2	9.0
59	B	$-NH_2$	p-Xylene	53)	>150°	—	—	>22	—
60	B	$-Cl$	HCB	55,56)	>180°	—	—	>23.5	—
61	B	$-Br$	HCB	55,56)	>140°	—	—	>21.5	—
62	B	$-SCH_3$	$CDCl_3$	57)	CH_3: −21° CH_2: − 3°	—	—	13.2 13.4	14.5 14.9
63	B	$-SC_6H_5$	CH_2Cl_2	57)	CH_3: −58° CH_2: −23°	—	—	11.9 12.4	13.0 13.7
64	B	$-SC(CH_3)_3$	$CDCl_3$	57)	CH_2: −25°	—	—	12.2	13.4
65	B	$-SCCl_3$	$CH_2Cl_2/CFCl_3$	57)	CH_2: −87°	—	—	9.1	10.0
66	B	$-As(CH_3)_2$	CH_2Cl_2	46)	CH_2: −73° CH_3: −76°	24 22	—	10.3 10.1	11.3 11.1
67	B	$-N=N-C_6H_4-pNO_2$	—	58)	CH_3: −62°	47	—	10.6	11.6

Table 3 (continued)

Com-pound No	Struc-tural Type	Substituent X on Nitrogen	Solvent[a]	Ref.	Literature data[a]			$\Delta G_c^{\neq[a]}$ kcal/mole	$\Delta H^{\neq[a]}$ ($\Delta S^{\neq} = +5$ eu) kcal/mole
					T_c (°C)	k_c (sec^{-1})	Other data		
68	B	$-Si(CH_3)_3$	CHF_2Cl	46)	$<-160°$	—	—	$\leqslant 5.5$ (?)	—
69	B	$-Si(OCH_3)_3$	CHF_2Cl	46)	$<-160°$	—	—	$\leqslant 5.5$ (?)	—
70	B	$-P(OCH_3)_2$	$CHFCl_2$	46)	$<-130°$	—	—	$\leqslant 6.7$ (?)	—
71	C	$-OCH_3$	p-Xylene	53)	$>130°$	—	—	>22	—
72	E	$-Br$	NB	51)	$125°$	1220	$E_a = 22.7$; log $A = 13$ ($\Delta H^{\neq} = 21.9$; $\Delta S^{\neq} = -1.7$)	17.9 (from k_c) (?) / 22.6 (from E_a) (?)	19.9(?) / 24.6(?)
73	E	$-F$	NB	51)	$>190°$	1890	—	>20.5	—
74b)	—	$-Cl$	CCl_4	59.60)	—	—	$k(80°) = 2.08 \times 10^{-4}$ / $k(110°) = 2.5 \times 10^{-3}$	26.7 / 27.1	28.5 / 29.0
75b)	—	$-Cl$	benzene	61)	—	—	$k(29.5°) = 4.3 \times 10^{-5}$	23.7	25.2
76	—	—	DCB	49)	$\sim150°$	140	—	~21	~23
77b)	—	—	C_2Cl_4	63)	—	—	$k(70°) = 1.33 \times 10^{-5}$	27.3 (70°)	29.0
78b)	—	—	C_2Cl_4	63)	—	—	—	27.1 (70°)	28.8
79b)	—	—	C_2Cl_4	63)	—	—	—	26.3 (70°)	28.0

80c)	—	—	C_2Cl_4	64	—	—	$k(100.5°) = 1.1 \times 10^6$; $\Delta H^{\neq} = 34.1$; $\Delta S^{\neq} = +5$ eu	—	34.1
81c)	—	—	C_2Cl_4	64)	—	—	$k(100.5°) = 7850 \times 10^6$; $\Delta H^{\neq} = 27.7$; $\Delta S^{\neq} = +6$	—	27.7
82c)	—	—	neat	46)	—	—	$\Delta G^{\neq}(120°) \sim 33$	—	~35
83b)	—	—	C_2Cl_4	65)	—	—	$\Delta G^{\neq}(115°) = 32.5$; $\Delta G^{\neq}(120°) = 32.4$	—	34.5
84b)	—	—	C_2Cl_4	65)	—	—	$\Delta G^{\neq}(115°) = 31.4$; $\Delta G^{\neq}(120°) = 31.4$	—	33.5

a) See footnotes b) and c) to Table 2.
b) Data from epimerization kinetics.
c) Data from racemization kinetics.

Table 4. *Activation parameters for pyramidal nitrogen inversion from NMR data: Four and five membered cyclic amines and derivatives*[a])

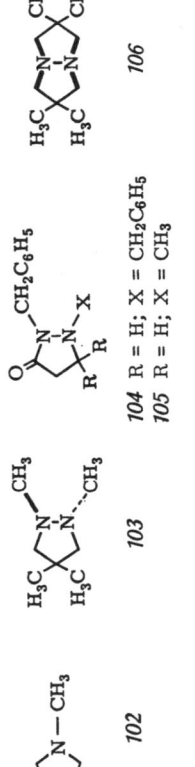

Compound No	Structural Type	Substituent X on Nitrogen	Solvent[a]	Ref.	Literature data[b] T_c (°C)	k_c (sec^{-1})	Other data	$\Delta G_c^{\neq a}$ kcal/mole	$\Delta H^{\neq a}$ ($\Delta S^{\neq} = +5$ eu) kcal/mole
85	F	—CH$_3$	CFCl$_3$	55)	CH$_2$: —105° / CH$_3$: —108°	35 / 32	— / —	8.2 / 8.3	9.0 / 9.1
86	F	—NH$_2$	CH$_2$Cl$_2$	46)	CH$_2$: —65° / CH$_3$: —74°	41 / 22	— / —	10.2 / 10.3	11.2 / 11.3
87	F	—ND$_2$	CFCl$_3$	46)	CH$_2$: —70° / CH$_3$: —82°	44 / 23	— / —	9.9 / 9.8	10.9 / 10.8
88	F	—NO$_2$	CFCl$_3$	46)	< —100°	—	—	<8	—
89	F	—Cl	CH$_2$Cl$_2$	55)	CH$_2$: —43° / CH$_3$: —47°	18 / 13	— / —	11.6 / 11.9	12.7 / 13.0
	F	—Cl	CFCl$_3$	55)	CH$_2$: —46° / CH$_3$: —56°	19 / 11	— / —	11.4 / 11.5	12.5 / 12.6
90	F	—Br	CH$_2$Cl$_2$	55)	CH$_3$: —56°	11	—	11.5	12.6
91	—	—N(—C$_6$H$_5$)	acetone	66)	—1°	—	—	13.3	14.7
92	G	—CH$_3$	CFCl$_3$	67)	—30°	2.10^3	—	10.3	11.5
93	G	—CH$_3$	CHFCl$_2$	68,62)	—107°	—	—	7.9	8.7
94	G	—Cl	CF$_2$Cl$_2$	69)	—64°	58	—	10.4	11.4

Table 4 (continued)

Compound No	Structural Type	Substituent X on Nitrogen	Solvent[a]	Ref.	Literature data[a] T_c (°C)	k_c (sec⁻¹)	Other data	$\Delta G^{\neq a}_c$ kcal/mole	$\Delta H^{\neq a}$ ($\Delta S^{\neq} = +5$ eu) kcal/mole
95	H	—CD$_3$	CHFCl$_2$	68)	—117°	—	—	7.4	8.2
96	H	—ND$_2$	CHFCl$_2$	68)	—98°	—	—	8.5	9.4
97	H	—OH	CDCl$_3$	68)	—11°	—	—	13.0	14.3
			D$_2$O	68)	+30°	—	—	15.0	16.5
98	H	—Br	CFCl$_3$	68)	—98	—	—	8.5	9.4
99	H	—Cl	CHFCl$_2$	68)	—79°	—	—	9.5	10.4
			CFCl$_3$	68)	—87°	—	—	9.0	9.9
100	—	—Cl	CFCl$_3$	68)	—98°	—	—	8.5	9.4
101	—	—Cl	CFCl$_3$	68)	—92°	—	—	8.8	9.7
102	—	—CH$_3$	CHFCl$_2$	68)	—100°	—	—	8.6	9.5
103	—	—N<	CH$_2$Cl$_2$	70)	—45°	—	—	11.1	12.2
104	—	—N<	Acetone-d$_6$	73)	—29°	—	—	11.7	12.8
105	—	—N<	Acetone-d$_6$	73	—49°	—	—	10.7	11.8
106	—	—N<	CDCl$_3$	79)	CH$_2$: —29°	—	—	12.1	13.3

107	I	—CH₃	CDCl₃	71)	42°	—		15.6	17.2
			D₂O	71)	62°	—		16.9	18.6
108	I	—CH(CH₃)₂	CH₂Cl₂	74)	+5°	15		14.8	16.2
109	I	—CH₂OCH₃	CH₂Cl₂	74)	−74°	34		10.3	11.3
110[b]	—	(—O)₂	CDBr₃	72)	—	—	$E_a = 29.2$; $\log A = 14.2$ ($\Delta H^{\neq} = 2.85$; $\Delta S^{\neq} = +3.9$)	27.0 (+100°)	28.7
111	—	—O<	CS₂	75)	−32°	4.3		13.3	14.5
			CDCl₃	75)	−21°	4.5		14.0	15.3
			CH₃OH/H₂O	75)	−18°	2.5		14.4	15.7

[a] See footnotes [b] and [c] of Table 2.
[b] Data from kinetics of racemization.

Table 5. *Activation parameters for nitrogen inversion from NMR data: Cyclic (> five membered), bicyclic and acyclic amines and derivatives[a])*

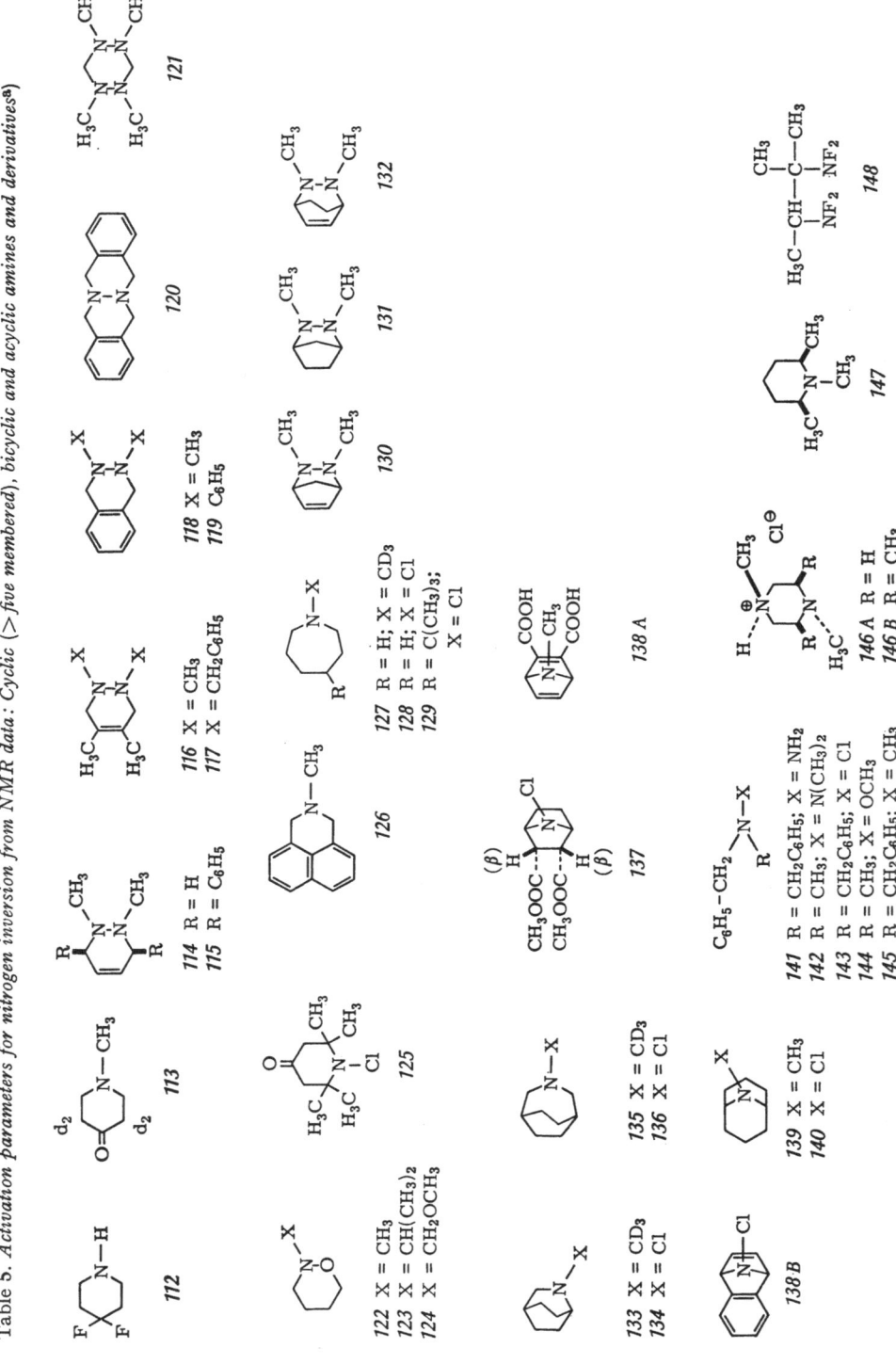

Compound No	Structural Type	Substituent X on Nitrogen	Solvent[a]	Ref.	T_c (°C)	k_c (sec^{-1})	Other data	$\Delta G_c^{\neq a}$ kcal/mole	$\Delta H^{\neq a}$ ($\Delta S^{\neq}=+5$ eu) kcal/mole
112	—	—H	Ac.	80)	−50°, −60°	780	$E_a = 10.5$	9.7	10.8
113	—	—CH₃	CHFCl₂	162)	−94°	126	—	8.6	9.5
114	—	—N<	CF₂Cl₂	76)	−20°	—	—	12.0	13.2
115	—	—N<	CF₂Cl₂	76)	−1°	—	—	13.3	14.7
116	—	—N<	CF₂Cl₂	76)	−14.5°	—	—	12.3	13.6
117	—	—N<	Py–d₅/Ac–d₆	78)	−27°	—	—	12.0	13.2
118	—	—N<	Py–d₅/Ac–d₆	78)	−27°	—	—	12.0	13.2
119	—	—N<	Py–d₅/Ac–d₆	78)	−23°	—	—	12.2	13.4
120	—	—N<	Py–d₅/Ac–d₆	78)	—	—	—	12.4	—
121	—	—N<	CP	81)	CH₂: −24°	—	—	11.8	13.0
122	—	—O—	Hexane	71a)	3°	—	—	13.7	15.1
	—	—|—	D₂O/CH₃OD (4/1)	71a)	33°	—	—	15.0	16.5
123	—	—O—	CH₂Cl₂	74)	−46°	8	—	12.5	13.6
124	—	—O—	CH₂Cl₂	74)	−51°	25	—	11.5	12.6

Table 5 (continued)

Compound No	Structural Type	Substituent X on Nitrogen	Solvent [a]	Ref.	Literature data [a] T_c (°C)	k_c (sec^{-1})	Other data	$\Delta G_c^{\neq}$ [a] kcal/mole	$\Delta H^{\neq}$ [a] ($\Delta S^{\neq} = +5$ eu) kcal/mole
125	—	—	CHFCl$_2$	46)	CH$_2$: $-106°$ / CH$_3$: $-115°$	92 / 17	— / —	8.1 / 8.2	9.0 / 9.0
126	—	—	CDCl$_3$ / D$_2$O/CD$_3$OD (1/3)	82) / 82)	$-72°$ / $-50°$	— / —	— / —	9.7 / 10.9	10.7 / 12.0
127	—	—CD$_3$ / —CH$_3$	CHF$_2$Cl / CHF$_2$Cl	162) / 62)	$-140°$ / $-125°$	36 / 44	— / —	6.4 / 7.3	6.9 / 8.0
128	—	—Cl	CHFCl$_2$	162)	$-100°$	40	—	8.4	9.3
129	—	—Cl	CHFCl$_2$	46)	$-90°$	39	—	8.9	9.8
130	—	—N$<$	Pentane	83)84)	$8°$	—	$\Delta H^{\neq} = 16.2$ $\Delta S^{\neq} = +7.6$	14.1	15.5
			CDCl$_3$	83)	$15.5°$	—	$\Delta H^{\neq} = 16.9$ $\Delta S^{\neq} = +8.3$	14.5	15.9
131	—	—N$<$	D$_2$O	83)	$47°$	—	—	16.1	17.7
		—N$<$	Pentane	83.84)	$-26°$	—	—	12.8	14.0
132	—	—N$<$	Pentane	83)	$-37.5°$	—	—	11.8	13.0
133	—	—CD$_3$	CHFCl$_2$	162)	$-98°$	56	—	8.4	9.3
134	—	—Cl	CHFCl$_2$	162)	$-54°$	47	—	10.6	11.7
135	—	—CD$_3$	CHFCl$_2$	162)	$-92°$	42	—	8.8	9.7

136	—	—Cl	CHFCl$_2$	162)	−65°	46	—	10.1	11.1
137	—	—Cl	HCB	46)	H(β): 140±10°; 70±10	—	—	21±1	23±1
138A	—	—CH$_3$	DMSO−d$_6$	157)	~96°	~22	—	19.5	21.3
138B	—	—Cl	CCl$_4$	87)	—	—	$k(23°) = 2.5 \times 10^{-5}$	23.5(23°)	25.0
139	—	—CH$_3$	CHFCl$_2$	46)	CH$_2$: −80±10°; 70±25	—	—	9.5±1	10.5±1
140	—	—Cl	CDCl$_3$	46)	CH$_2$: 0±10°; 70±25	—	—	14±1	15.5±1
141	—	—NH$_2$	CFCl$_3$/CH$_2$Cl$_2$	85)	−95°	—	—	8.5	9.4
142	—	—N(CH$_3$)$_2$	CF$_2$Cl$_2$	92)	CH$_2$: −130°	135	—	6.8	7.4
143	—	—Cl	CHFCl$_2$	46)	−75°	40	—	9.6	10.6
144	—	—OCH$_3$	Hexane	86)	−16°	35	$E_a = 12.9$; $\log A = 12.9$ ($\Delta H^{\neq} = 12.3$; $\Delta S^{\neq} = -1.7$)	13.0	14.3
145	—	—CH$_3$	VC	166)	−137°	86		6.5	7.2
146A[b]	—	—CH$_3$	H+/H$_2$O[b]	28)	—	—	$k(25° = 2 \times 10^5$	8.4(25°)	9.9
146B[b]	—	—CH$_3$	H+/H$_2$O	88)	—	—	$k(44°) = 2 \times 10^5$	7.7(44°)	9.3
147[b]	—	—CH$_3$	H+/H$_2$O	89)	—	—	$k(33°) = 1.5 \times 10^4$	9.3(33°)	10.8
	—	—CH$_3$	H+/H$_2$O	90)	—	—	$k(33°) = 500$	11.0(33°)	12.5
148	—	—	NB	94)	>147°	>3600	—	>18.0	>20.1

a) See footnotes [b] and [c] to Table 2.
b) Results obtaining using the method of Saunders and Yamada [28].

upon the nitrogen site. It is seen from the data in Tables 1—5 that including the nitrogen site in a three-membered ring $(\theta \sim 60°)$ strongly hinders the opening of θ, destabilizes the TS with respect to the pyramidal form and leads to a very marked increase in inversion barrier (of the order of 15 kcal/mole) as compared for instance to the corresponding five-membered cyclic systems $(\theta \sim 105—110°)$ (see: 32 and 95; 77 and 103; 80 and 107; 74 and 99; see also: aziridine, 16, and dimethylamine 9). The ring strain effect is more pronounced for the diazirine (77), oxaziridine (80) and N-chloroaziridine (74) systems (ca. 17 kcal/mole barrier increase), where the barrier to inversion is higher, than for aziridines (18; 32) (ca. 11 kcal/mole increase). The strain effect of a four-membered ring is much smaller $(\theta \sim 90—95°)$ (see 85; 91; 89), and amounts to ca. 1 or 2 kcal/mole with respect to the corresponding five-membered ring systems.

The inversion barrier increases again in the six-membered ring systems (112, 113, 114) and then decreases in the seven-membered rings (127, 128). In these cases, although angle strain also plays a role (slight C—N—C angle opening is expected in seven-membered cyclic systems [95,97]) torsional strain is expected to contribute to the barrier height.

In the series of N—CH_3 cyclic tertiary amines the relative order of barrier heights as a function of ring size is:

n=3 (17; 32) >n=6 (113)~n=4 (85) >n=5 (93, 95) >n=7 (127)~ acyclic (145).

Barriers of ca. 6—7 kcal/mole may be expected for larger ring systems.

In the case of *bicyclic systems (133—140)* there are not yet enough data available for a detailed discussion to be given. The barriers obtained for the bicyclo [2.2.2]octane (133, 134) and the bicyclo[3.3.1]nonane (139, 140) compounds are in the range expected for the corresponding monocyclic systems. In the case of the bicyclo[3.2.2]nonane derivatives 135 and 136 the inversion barriers are larger than in the seven-membered ring compounds 127 and 128.

The most remarkable effect is found for 7-aza-bicyclo[2.2.1]heptane compounds (137, 138 A, 138 B). The barriers measured in these systems are comparable to those found in the corresponding aziridines and are thus much higher than expected on the basis of angle strain $(\theta \sim 95°$ in 137, 138) [c]. The origin of this "bicyclic effect" is not yet clear, but an important contribution may arise from destabilization of the TS by repulsions between the nitrogen lone pair and the bonding electrons in both two carbon bridges of the bicyclic system.

---

[c] The C(1)—C(7)—C(4) angle is equal to 96° in norbornane [98].

3.2. Electronic Effects

In the present section we shall try to rationalize the variations in inversion barrier in terms of a certain number of electronic effects characterizing the modes of electronic structure changes brought about especially by atoms or groups directly linked to the inverting nitrogen site. In many cases, different effects may be operative in the same compound and their relative contributions are difficult to separate. A discussion of the relation between such effects and more detailed theoretical treatments will be given below (Section 5).

A. $(p-p)\pi$ Conjugation

Conjugation of the nitrogen lone pair with an adjacent π-system is greater in the planar TS, where the lone pair is in a pure p orbital, than in the pyramidal state, where the lone pair orbital is s, p hybridized. The presence of such conjugative interactions is expected to lead to two types of changes: a geometrical change: the nitrogen site becomes less pyramidal; and an energetic change: the nitrogen inversion barrier decreases. For instance, in the case of formamide *13*, the conjugation of the nitrogen lone pair with the carbonyl group flattens the nitrogen pyramid ($\varphi\sim 17°$; [35]) and decreases the inversion barrier (1.1 kcal/mole) with respect to ammonia ($\varphi = 61°$; 5.8 kcal/mole). The same holds for cyanamide *11*, nitramine *14*, and aniline *10* where the nitrogen inversion barrier decreases as the nitrogen site becomes more planar (*11*, $\varphi\sim 38°$; *14*, $\varphi\sim 51°$; *10*, $\varphi\sim 39°$ [32–34]). Microwave data are in agreement with a planar nitrogen site in pyrrole [99].

Although the various groups capable of $(p-p)\pi$ conjugation are of different size and thus lead to different steric effects, the influence of conjugation in lowering the inversion barrier is beyond doubt, the amount of decrease depending both on the conjugative ability and on the size of the group.

The tables show examples for the following conjugative substituents: $-C_6H_5$ *(10, 27, 36–41, 52)*, $-COR$ *(13, 28–31, 42)*, $-CN$ *(11)*, $-NO_2$ *(14, 88)*.

An interesting case is that of $H_2N-CO-O-NH_2$ where one nitrogen site is found to be very nearly planar (H_2N-CO) whereas the other one is strongly pyramidal ($O-NH_2$) [100].

In the case of aziridines, the angle strain of the cyclic system opposes the tendency of conjugative groups to render the nitrogen site less pyramidal. The pyramidality angle φ is found to be of ca. 50°—60° in N-benzoylaziridine derivatives [101]. The weakening of carbonyl-nitrogen lone-pair conjugation in $N-CO-R$ aziridines [43,102] leads both to lower barriers to rotation about the $N-CO$ bond (< 6 kcal/mole; [43]) and to higher nitrogen inversion barriers (*30, 31* <6 kcal/mole; *28*, 7.8 kcal/mole) than in acyclic amides (rotation barriers of ca. 20 kcal/mole [17b]); inversion barriers presumably of the order of 1 or 2 kcal/mole, see *13*).

The N—COOCH$_3$ *(28)* and N—CON(CH$_3$)$_2$ *(29)* aziridines display appreciably higher barriers to inversion than the N-acyl aziridines *(30, 31)*. The conjugative ability of the —OCH$_3$ and —N(CH$_3$)$_2$ groups decreases the conjugation between the aziridine nitrogen and the carbonyl group, thus increasing the barrier to nitrogen inversion; the effect is larger for —N(CH$_3$)$_2$ which is a better conjugative group than —OCH$_3$.

In the homogeneous series of aziridines *36—41* where the nitrogen site bears an —C$_6$H$_4$—Y group, one observes a decrease in inversion barrier which parallels the ability of the substituent Y to withdraw electron density from the aromatic ring. The barrier decreases in the order:

$$Y = p—OCH_3 \; > \; —H \; > \; p—Cl \; > \; m—CF_3 \; > \; p—CF_3 \; > \; p—NO_2 \text{ [47]}$$

B. (d—p)π Conjugation [d]

When elements with low-lying *d* orbitals are directly attached to the nitrogen site, it is possible to imagine conjugation of the lone pair with an empty *d* orbital ((d—p)π overlap). As for (p—p)π conjugation, the effect is expected to be more pronounced in the planar TS than in the pyramidal configuration, leading to a decrease in inversion barrier. Furthermore, (d—p)π conjugation should be more pronounced for second-row elements (Si, P, S, Cl) than for third or higher row elements (As, Se, Br . . .) as the overlap of the 2*p* lone pair with a 3*d* orbital is larger than with a 4*d* orbital.

Aziridines bearing Si *(68, 69)*, P *(58, 70)*, S *(55—57, 62—65)*, As *(66)* on nitrogen display low inversion barriers, especially in the case of Si and P [e].

It has also been found that the nitrogen site is planar in silylamines, e.g. (SiH$_3$)$_2$NH, [Si(CH$_3$)$_3$]$_2$NH, (SiH$_3$)$_2$NCH$_3$ [105]; see also [77] [f].

Thus the nitrogen inversion barriers are found to be small and the nitrogen sites are planar in those compounds (especially N—Si) where (d—p)π conjugation might be important. There is, however, no indication about the existence and the effect of such conjugation, and a separation from other factors, which may also account for the low barriers (especially inductive electron release by the silicon atom), is not possible at this stage (see the theoretical study of H$_2$N—SiH$_3$; section 5.4B).

---

[d] The role of *d* orbitals is also discussed in Section 5.

[e] For an IR study of conjugation effects in N—P aziridines see [103]. Nitrogen inversion seems also to be fast in N—Sn aziridines [104].

[f] Trigermylamine (GeH$_3$)$_3$N is non-planar [106].

C. Effect of Heteroatoms

Substituent electronegativity. Electron repulsion effects

The effects of substituent electronegativity, electrostatic and electron repulsion interactions are very difficult to separate as electronegative substituents also bear lone pairs which lead to repulsive interactions.

Within the hybridization scheme, electronegative substituents on nitrogen are expected to increase the s character of the lone pair orbital [96]. As the lone pair is in a p orbital in the transition state, such substituents should increase the barrier to nitrogen inversion.

The electronegativity value of a given substituent is not an accurate parameter; however, one may characterize a substituent by the electronegativity of the central atom or of the whole group as estimated in the literature [107-109]. Fig. 4 represents a plot of barrier heights versus electronegativity for a series of compounds. It should be unnecessary to stress the *crudeness* of such a plot, but some special features seem to emerge.

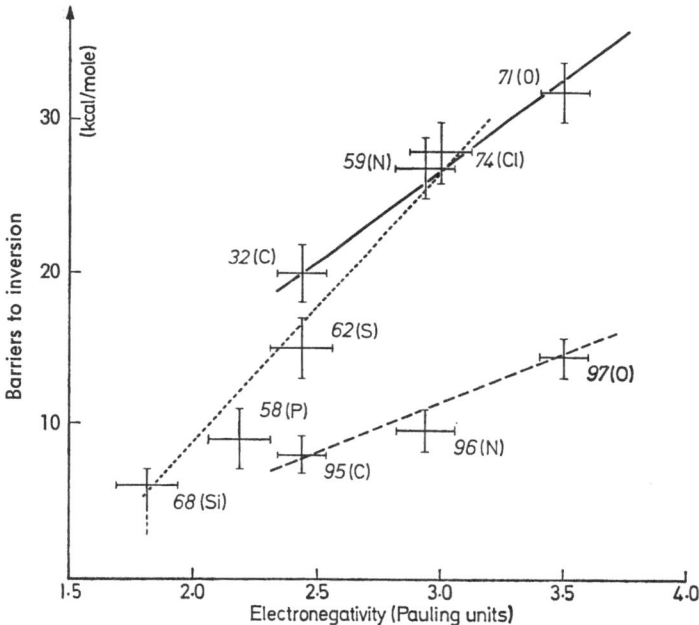

Fig. 4. Plot of barriers to nitrogen inversion versus the Pauling electronegativity index of various substituents linked to nitrogen. The compound taken as reference is indicated by its number (see Tables 2—5) and the changing substituent is given in parentheses. One could also have used the sum of the electronegativities of all three substituents on nitrogen

There is a marked barrier increase along the series $N-CH_3$, $N-NH_2$, $N-OR$ in a series of aziridines [g] *(32, 59, 71)* and in pyrrolidines *(95, 96, 97)* as substituent electronegativity increases.

Such effects have been observed in a great variety of monocyclic, bicyclic and acyclic N—chloro amines, hydrazine and hydroxylamine derivatives (see Tables 2—5). When two electronegative substituents are attached to the same nitrogen site a very large barrier increase is expected; this is the case for the N-methoxy-1,2-oxazolidine *110* and for the N-difluoroamine *148* where the presence of two oxygen or two fluorine atoms leads to barriers of ca. 29 kcal/mole and $\geqslant 20$ kcal/mole respectively. A barrier of ca 60 kcal/mole has been estimated for NF$_3$ [140]. Such very high barriers are also found in N, O, Cl bearing aziridines where both electronegativity and strain effects are operative (Table 3). In the case of second-row substituents (Cl: *74*; SCH$_3$: *62*; $-P(O)(C_6H_5)_2$: *58*; $-Si(CH_3)_3$: *68*) (Fig. 4), the barrier appears to be somewhat lower than for first-row substituents of similar electronegativity especially at low electronegativities and except for chlorine. Such a "second-row effect" might arise from $(d-p)\pi$ overlap, but could also be an artefact due to the crudeness of the plot.

In addition to polarity effects due to electronegativity differences, *polarisability* should also be considered. The larger polarisability of second row substituents and of bonds with second row atoms may contribute to the occurence of a specific "second row effect".

Substituents bearing electronic lone pairs may also increase the inversion barrier through *electron repulsion interactions* with the nitrogen lone pair, which are higher in the TS than in the GS. On the other hand, electrostatic *dipole-dipole interactions* are expected to decrease in the TS where the contribution of the nitrogen lone pair to the local dipole moment vanishes.

It is also necessary to consider the electrostatic nuclear-electron attraction or nuclear-nuclear repulsion effects brought about by the various substituents and which may differ greatly from one substituent to another.

Finally, steric effects also contribute. Thus it seems wiser at this stage to consider empirical substituent effects as an unseparable whole. Relationships like the barrier-electronegativity "correlation" shown in Fig. 4 are at best crude empirical leads without physical explanatory power, but which may nevertheless be of some pratical use.

---

[g] The inversion barriers for compounds *59* (ca. 26—27 kcal/mole) and *71* (ca. 30—32 kcal/mole) have been estimated from the values obtained for diaziridines and oxaziridines (Table 3) and from a study of substituent effects in pyrrolidines [68]. From Fig. 4 barriers of ca. 40 and 18—20 kcal/mole may be estimated for N-fluoro aziridines and N-fluoro pyrrolidines respectively.

When one considers such total substituent effects for different types of structure, it is found that there are *structural effects on substituent effects*, i.e. the change in barrier brought about by a given substituent with respect to the parent tertiary amine depends on the molecular structure. Thus, the barrier-raising effect of N, O, Cl heteroatoms is ca. 6 kcal/mole higher in aziridines *(32, 77, 83, 74)* than in the corresponding pyrrolidines *(95, 103, 107, 99)* [68] and slightly higher (ca. 1 kcal/mole) in azetidines *(85, 86, 89)* than in pyrrolidines. It is seen that the effect of replacing a N—CH$_3$ by a N—Cl group varies markedly from one type of structure to another (see for instance *32, 74; 85, 89; 95, 99; 127, 128; 133, 134; 135, 136*).

There is presently not much evidence for the influence of heteroatoms not directly linked to the inverting nitrogen except for fluoroalkyl amines. The presence of α fluorine atoms, as in N—C—F groups, leads to a decrease in inversion barrier (see *47—50, 53*).

Stabilization of the TS by bond-no bond resonance (C—F hyperconjugation) has been suggested for explaining this effect. However, a study of carbanion stabilization by fluoroalkyl groups favours an inductive field effect over fluorine hyperconjugation [110]. Various dipole-dipole interactions may also contribute to the barrier decrease. Thus the barrier-lowering effect of fluoro alkyl groups might arise at least in part from a destabilization of the GS. Similar barrier-lowering effects seem to be present in compounds containing N—C—O groups *(24; 109)*. Nitrogen sites coordinated to transition metals have been found to invert slowly, the corresponding barriers being of the order of 10—15 kcal/mole [91].

D. Tentative Generalization

Using the results discussed above one may try to draw a schematic picture of the effect of substituents on nitrogen inversion barriers in terms of the electronic redistributions they are producing. This picture and the discussed electronic factors should be considered as *ad hoc* reasoning based on the "effects" described above, rather than as a real physical explanation.

One may distinguish π effects and σ effects concerning respectively the nitrogen lone pair and the three N—X σ bonds. π effects have a more pronounced influence on the TS than on the GS, the reverse holding for σ effects.

π effects:

— delocalization of the nitrogen lone pair by conjugation diminishes the lone pair electron density and leads to a barrier *decrease*, through stabilization of the TS;

— accumulation or localization of electron density in the nitrogen lone pair by electron repulsion destabilizes the TS more than the GS and leads to a barrier *increase*. (This type of effect has also been described as "conjugative destabilization" [72]. For a discussion of electron pair repulsion see also ref. [111]).

σ effects:

— electron attraction away from nitrogen along a σ bond *increases* the inversion barrier[h];
— electron donation towards the nitrogen site along a σ bond *decreases* the inversion barrier[i].

These effects may be summarized as follows:

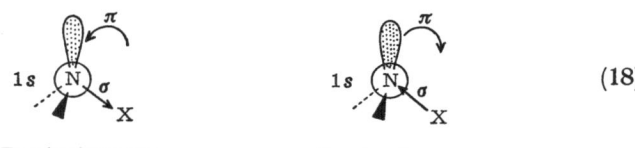

Barrier increase   Barrier decrease

$$(18)$$

These effects could also be formulated within the hybridization scheme:

Nitrogen inversion represents a $sp^n \rightarrow p$ promotion of the nitrogen lone-pair electrons. Effects which increase s character in the GS (electronegative substituents, also angle strain [96]) increase the inversion barrier (and conversely); effects which increase the p character in the GS

---

[h] This effect may be ascribed to a relative stabilization of the GS with respect to the TS. As the GS is more compact than the TS, electronic repulsions between the three σ bonds are larger, so that an electron-attractive substituent, which decreases the electron density in the bonds and in the vicinity of the nitrogen atom, produces a larger decrease in these bond-bond repulsions in the GS than in the TS. One might then even try to "rationalize" the larger barrier increase produced by a given electro-attractive substituent in aziridines as compared to pyrrolidines, for instance: aziridines being more compact molecules because of the smaller endocyclic angle, one might expect that introducing an electro-attractive substituent both stabilizes the GS more and stabilizes the TS less than in pyrrolidines. Interactions between the bonding electrons and the $1s$ electrons on nitrogen are also expected to favour the GS with respect to the TS where the nitrogen site becomes more electronegative (sp^2) and attracts more electron density along the σ bonds, thus leading to increased $1s$ (on N)-bond repulsions (see also discussions in section 5.4.B).

[i] In this case accumulation of electron density in the σ bonds leads to greater electronic repulsions (bond-bond, $1s$(N)-bond) in the pyramidal state, which is therefore destabilized with respect to the TS. This is an alternative way (to be preferred to $(d-p)\pi$ bonding; see section 5.4.B) of accounting for the low barriers found for nitrogen sites bearing electron releasing second-row substituents (for instance, N-silylamines).

(conjugation; angle opening) decrease the inversion barrier (and conversely).

3.3. Distinction of Nitrogen Inversion from Other Rate Processes

As pointed out in Section 2, there are cases where both nitrogen inversion and another process, ring inversion or internal rotation, may occur together in the same molecule. It then becomes difficult to decide which process is being observed. If the two processes correspond to two distinct transition states they may, in principle, be observed separately (for instance, two different types of coalescence in NMR spectra). The two processes may also occur together by an intermediate mechanism, through a *single* transition state in the potential surface; then the question of distinguishing them loses its meaning.

In order to resolve the ambiguity one may make use of well-chosen structural effects which modify in a known fashion the barrier of one of the possible processes. Thus, if one has to distinguish between nitrogen inversion in R_2NX and internal rotation about the R_2N-X bond, one may for instance use the following procedures:

1. introducing the nitrogen site into a strained system (e.g. a three-membered ring),

2. conjugating the nitrogen with one or two carbonyl (or other) groups (as in amide or succinimide derivatives).

3. increasing the bulk of the substituents on nitrogen.

If the process which is being observed (generally by NMR spectroscopy) is nitrogen inversion, procedure 1 should increase markedly the inversion barrier, 2 and 3 should decrease it (the barrier should even disappear in succinimide derivatives). These procedures have been used for instance for distinguishing between nitrogen inversion and hindered rotation about the $N-O$, $N-S$ and $N-P$ bonds in hydroxylamines [112], sulfenamides [57,113] and amino phosphines [114].

In the case of the acyclic hydrazine *141, 142* and hydroxylamine *144* derivatives, both nitrogen inversion and hindered rotation is present. The results have been interpreted in terms of nitrogen inversion [85,86,92] in agreement with the magnitude of the substituent effects in the series *141, 143—145* which are similar to those obtained for the corresponding pyrrolidine derivatives *(95—97, 99)*. However, steric effects would favour hindered rotation about the $N-O$ bond as the observed (NMR) process in acyclic hydroxylamines [112].

Ring inversion may occur together with nitrogen inversion. The barriers to inversion of the four-membered ring in azetidines and of the five-membered ring in pyrrolidines are very low (1.3 kcal/mole and less than 0.5 kcal/mole respectively [115]) so that ring inversion is much faster than nitrogen inversion, which is thus the process observed by NMR.

In the case of piperidine derivatives, the identification of the process is difficult, as inversion of the six-membered ring and nitrogen inversion are expected to have similar barriers [69]. The ambiguity can be removed by introducing an sp^2 carbon in the ring (carbonyl or $C=CH_2$ group) as in 4-piperidones *(113; 125)* [46]. The barrier to ring inversion is then expected to be much lower (cyclohexanone: < 5.2 kcal/mole; methylene cyclohexane: 7.7 kcal/mole [116]) and the process observed in compounds *113* and *125* is nitrogen inversion. The same is true for cyclohexene type compounds and, for instance, both nitrogen inversion (barrier of 12.3 kcal/mole) and ring inversion (barrier of 8.2 kcal/mole) have been observed for compound *116* [76].

Because of pseudorotation, inversion of the seven-membered ring is very fast (barrier of the order of 2 kcal/mole [97,117]) so that the rate process observed in compounds *127—129* must be nitrogen inversion. In eight-membered cyclic amines the nitrogen inversion barrier is expected to be of the same order as the ring inversion barrier (7.7 kcal/mole [118]); however, in this case introducing a carbonyl group does not lead to a marked decrease in ring inversion barrier [119]. The ring inversion-nitrogen inversion ambiguity in the tetrahydro-1,2-oxazines *122—124* is not entirely resolved although the results presently available favour nitrogen inversion as the observed process.

3.4. Structural Effects on Planar Nitrogen Inversion

Table 6 gives a list of selected data on planar nitrogen inversion, the process depicted in Eq. (2) (Section 1 above), by which two bent sp^2 type nitrogen sites interconvert through a linear transition state.

A. Planar Nitrogen Inversion Versus Rotation about the C=N bond

Before discussing structural effects on barrier heights, it is necessary to distinguish between the two processes: planar nitrogen inversion and rotation about the C=N bond, which may both lead to interconversion of the isomers. Rotation about a non-activated C=N double bond is expected to be hindered by an energy barrier similar to the ethylene barrier, i. e. of the order of 50—60 kcal/mole [121,132]. Thus, alkyl- and presumably also aryl-substituted imines, which show interconversion barriers below 30 kcal/mole (Table 6), undergo nitrogen inversion[j].

The situation is less clear when heteroatoms are present which may perturb appreciably the C=N bond and eventually bring the barrier to rotation below the barrier to inversion.

---

[j] This is also supported by calculations on $CH_2=NH$ (see Section 5).

The main procedure used for deciding whether rotation or inversion is occurring, consists in studying structural effects on the barrier height and comparing them with similar effects on pyramidal nitrogen inversion. It rests on the reasonable hypothesis that both pyramidal and planar inversion processes should show a similar dependence on structural factors. Medium effects have also been considered [125]. Detailed studies of substituent effects in alkyl- or aryl-substituted [120-122] as well as in heteroatom-bearing [121,123-126] imines display strong analogies with results obtained for pyramidal nitrogen inversion. It thus seems reasonable to admit that the observed process is probably planar nitrogen inversion, although in specific cases important perturbations by substituent effects may render rotation about the C=N bond easier than nitrogen inversion or may lead to a single TS in the rotation-inversion potential energy surface. In carbodiimides (see *181* for instance), both barriers are calculated to be of similar energies [165].

B. Steric Effects

Steric effects are greater in the bent GS than in the linear TS and should lead to a decrease in inversion barrier as for pyramidal inversion.

A barrier decrease is indeed observed in a series of o,o'-substituted sterically crowded acetone-anils [120] as the size of the substituents increases. Similarly, increasing substituent size decreases the isomerization barrier in o,o'-disubstituted N-aryl-guanidines; in contrast, a barrier increase is observed in the corresponding salts where only rotation about the C=N$^+$ bond is possible [124]. However, such steric effects do not seem to be conclusive. They are complicated by the fact that o,o'-disubstituted N-aryl compounds act also on the conjugation between the nitrogen site and the aryl group.

C. Electronic Effects

In N-aryl compounds YZC=NAr (for instance, types *J*, *K*, *L*, Table 6) bearing two substituents Y, Z on the imino carbon, the plane of the aryl group is probably rotated with respect to the C=N—C plane so as to relieve steric interactions[k]. Such a conformation permits $(p-p)\pi$ conjugation of the nitrogen lone pair with the aromatic π system.

Indeed, replacing N—C(sp^3) *(149, 151, 166)* by N—C(aryl) *(150, 152, 167)* leads to a decrease in energy barrier by at least 5 kcal/mole. Extensive studies of substituent effects in *p*-substituted N-phenyl systems of type *J (154, 155*; ref. [121]), *K (157—161*; ref. [122]) and *L (169—174*; ref. [124, 126]) show that the effects are analogous in the three series, indicating

---

[k] The aniline ring in $C_6H_5-N=CHC_6H_5$ is twisted out of the C—N=C—C plane by 40—55° (X-Ray analysis) [133].

Table 6. *Activation parameters for planar nitrogen inversion from NMR and isomerization data*[a]

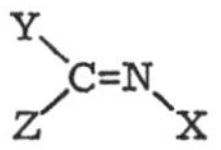

$$Y\!-\!\underset{Z}{\overset{}{C}}\!=\!N\!-\!X$$

J

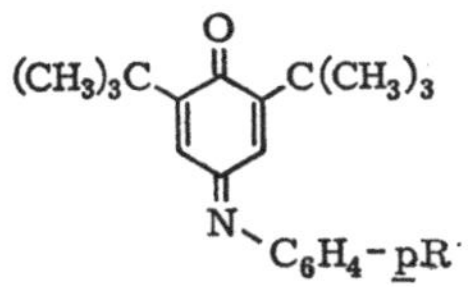

$$(CH_3)_3C\!-\!\overset{O}{\underset{C(CH_3)_3}{\bigcirc}}\!-\!N\!-\!C_6H_4\!-\!\underline{p}R'$$

K

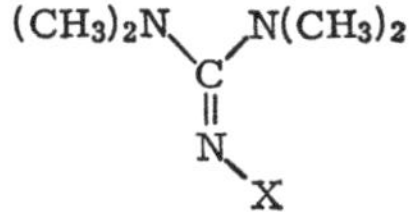

$$(CH_3)_2N\!-\!\underset{N\!-\!X}{\overset{}{C}}\!=\!N\!-\!X$$

$$L\ \ X = C_6H_4\text{-}\underline{p}R$$
M

$$F\!-\!\underset{F}{\overset{CF(CF_3)_2}{C}}\!=\!N\!-\!X$$

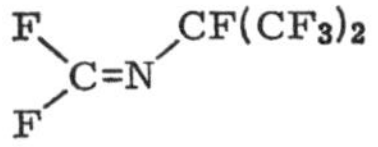

180

$$(CH_3)_2CH\!-\!N\!=\!C\!=\!N\!-\!CH(CH_3)_2$$

181

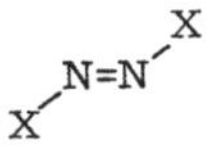

$$\underset{X}{\overset{X}{N\!-\!N}}\!=\!N\!-\!X$$

182 X = F
183 X = C_6H_5

Com-pound No	Struc-tural Type	Substituent	Solvent[a,b]	Ref.	Literature data[a]			$\Delta G_c^{\neq}$[a] kcal/mole	$\Delta H^{\neq}$[a] ($\Delta S^{\neq} = +5$ eu) kcal/mole
					T_c (°C)	k_c (sec^{-1})	Other data		
149	J	X = -CH RAr, Y = Z=CH$_3$	Q	120)	>180°	—	—	>23	>25
150	J	X = -C$_6$H$_5$, Y = Z=CH$_3$	DPE	120)	126°	56	—	20.3	22.3
151	J	X = -CH$_3$, Y = -C$_6$H$_4$A-p, Z = -C$_6$H$_4$A'-p	CH	121)	—	—	$k(60°) = 10^{-4}$	—	$E_a = 25\text{--}27$

No.		Substituents	Solvent	Ref	angle				$E_a = 17$–20
152	J	X = —C_6H_4A—p Y = —C_6H_4B—p Z = —C_6H_4B'—p	CH CCl_4	121)	—	—	A=$N(CH_3)_2$: $E_a = 19.7$ $\Delta S^{\neq} = -2$	—	—
153	J	X = Br Y = —C_6H_4Cl—p Z = —C_6H_5	CH	121)	—	—	$k(60°)<10^{-6}$	>28(60°)	—
154	J	X = Cl Y = —C_6H_4Cl—p Z = —C_6H_5	CH	121)	—	—	$k(60°)<10^{-8}$	>31(60°)	—
155	J	X = OCH_3 Y = —C_6H_4Cl—p Z = —C_6H_5	decane	121)	—	—	$k(60°)<10^{-13}$	>39(60°)	—
156	J	X = —CN Y = Z = —CH_3	Ac	128)	85°	19	—	18.9	20.7
157	K	R = —$N(CH_3)_2$	TCB	122)	144°	10	—	22.8	24.9
158	K	R = —OCH_3	TCB	122)	152°	15	—	22.9	25.0
159	K	R = —H	TCB	122)	140°	20	—	22.0	24.1
160	K	R = —$COCH_3$	TCB	122)	90°	21	—	19.2	21.0
161	K	R = —NO_2	TCB	122)	68°	21	—	18.0	19.7
162	J	X = —$CH_2C_6H_5$ Y = Z = —OCH_3	Ac	127)	74°	5.5	—	19.3	21.0
163	J	X = —C_6H_5 Y = Z = —OCH_3	Ac	123.127)	0°	20	—	14.3	15.7

353

Table 6 (continued)

Compound No	Structural Type	Substituent	Solvent[a,b]	Ref.	Literature data[a] T_c (°C)	k_c (sec⁻¹)	Other data	$\Delta G_c^{\neq a}$ kcal/mole	$\Delta H^{\neq a}$ ($\Delta S^{\neq} = +5$ eu) kcal/mole
164	J	X = —Cl, Y = Z = —OCH₃	DPE	123,127)	>160°	>13.3	—	>23	>25
165	J	X = —CN, Y = Z = —OCH₃	Ac	128)	—16°	5	—	14.1	15.5
166	J	X = —CH₃, Y = Z = —SCH₃	DPE	123)	73°	13.3	—	18.6	20.3
167	J	X = —C₆H₅, Y = Z = —SCH₃	Ac	123)	—22°	5.5	—	13.7	15.0
168	J	X = —CN, Y = Z = —SCH₃	Ac	123,128)	—1°	29	—	14.0	15.4
169	L	R = —N(CH₃)₂	CH₃OH	124)	—2°	53	—	13.7	15.1
170	L	R = —OCH₃	CH₃OH	124)	0°	53	—	13.8	15.2
171	L	R = —CH₃	CH₃OH	124)	—12°	53	—	13.1	14.4
172	L	R=H	CH₃OH CDCl₃	124) 124,126)	—17° —35°	55 —	— —	12.9 12.1	14.2 13.3
173	L	R = —COCH₃	CH₃OH	124)	—68°	51	—	10.2	11.2
174	L	R = —NO₂	CH₃OH	124)	—85°	40	—	9.5	10.4
175	M	X = —N(CH₃)₂	TCB	125)	116°	—	—	21.1	23.0

No.		X	Solvent	Ref.					
176	M	X = —OCH₃	TCB DCB	125) 129)	105° 152°	— —	— —	19.9 22.1	21.8 24.2
177	M	X = —CH₃	CDCl₃	125,129)	73°	—	—	18.8	20.5
178	M	X = —CH(CH₃)₂	CDCl₃	125)	53°	—	—	17.5	19.1
179	J	X = —CN Y = Z = —NDCH₃	Ac	128)	—43°	11	—	12.3	13.5
180	—	—	CFCl₃	130,135)	—	—	$k(63°) = 5 \cdot 10^3$	14.7(63°)	16.4
181	—	—	VC, CHFCl₂; 1/1	131)	—140°	13	—	6.7	7.4
182	—	—	Gas phase	132)	—	—	$E_a = 32$	—	—
183	—	—	—	142)	—	—	$E_a \sim 23$	—	—

a) See footnotes b) and c) to Table 2.
b) Q: quinoline; DPE: diphenylether; CH: cyclohexane; DCB: 1,2-dichlorobenzene; TCB: 1,2,4-trichlorobenzene.

that the isomerization process should be the same. In addition these effects are very similar to those observed for a series of *p*-substituted N-phenyl aziridines *(36—41*; Table 2; ref. [47])*;* for instance, replacing $N-C_6H_4-p$OCH$_3$ by $N-C_6H_4-p$NO$_2$ leads to a ca. 4.5 kcal/mole barrier decrease in series *K (158, 161)* and *L (170, 174)* as well as in the aziridines *36* and *40* (Table 2)[1].

It has also been shown that the isomerization rates of *para*-substituted C-aryl-N-aryl imines show a more pronounced dependence on the *para* substituent in the N-aryl ring than on the *para* C-aryl substituent; this again favours nitrogen inversion as the operative process [121,134].

Electronegative heteroatomic substituents on nitrogen lead to a very marked increase in isomerization barrier, as has been found in N-halo imines *(153, 154, 164)* and in oximes *(155)*. A N-fluoroalkyl group *(180)* decreases the inversion barrier.

Thus both steric and electronic effects on planar nitrogen inversion parallel the behavior observed for pyramidal nitrogen inversion. The picture given above (section 3.2.D) for this last process should presumably also apply to the present case. (See also below the analogies between the theoretical picture of pyramidal and planar nitrogen inversion.)

4. Medium Effects on Nitrogen Inversion

The nature of the medium may also affect the height of a nitrogen inversion barrier. One may grossly distinguish two types of effects:

1. as the GS is more polar than the TS, increasing solvent polarity should stabilize the GS and increase the inversion barrier;

2. in solvents capable of hydrogen bonding, the formation of a hydrogen bond to the nitrogen lone pair should also stabilize the GS and increase the inversion barrier.

Both effects 1 and 2 may be more or less obscured by the presence of heteroatoms on nitrogen or other polar groups which may also present solvent-dependent interactions. For instance when different rotamers are present, changing solvent may also affect the rotamer populations (or the equilibrium geometry of a single rotamer) and thus modify the inversion barrier. Such effects may occur, for instance, in hydroxylamine [86] and hydrazine [76] derivatives.

Changing from a non-polar (or slightly polar) to a hydroxylic solvent leads to a barrier increase of about 2 kcal/mole *(18, 46, 97, 107, 111,*

---

[1] One might argue that, if the aryl group and the YZC=N group were coplanar in YZC=N—Ar (so as to permit conjugation between the C=N π bond and the aryl system), one could expect similar substituent effects on the C=N rotation barrier (especially for compounds of type *L*).

122, 130). Water itself seems to produce a somewhat larger effect than methanol. As a consequence the inversion barriers determined in water solution by the method of Saunders and Yamada [28,88—90] *(145—147 B)* probably incorporate such a solvent effect. It has been found that the rate of nitrogen inversion in the bicyclic hydrazine *130* in water solution is strongly pH-dependent and increases when the pH of the solution is lowered [76].

Deuterochloroform solutions seem to yield slightly higher (ca. 0.5 kcal/mole) barriers than solutions in hydrocarbon solvents (see for instance *111, 139*; [83]).

A similar increase is observed from $CFCl_3$ to $CHFCl_2$ solutions (*99*; [68]). Such solvent effects have been found in planar nitrogen inversion processes as well [124,125].

They have been used in some cases for providing additional evidence in distinguishing nitrogen inversion from other rate processes (internal rotation, ring inversion), the idea being that these other processes are probably less sensitive to a change from non-hydroxylic to hydroxylic solvent (for instance, hexane to water) than nitrogen inversion itself (see for instance [71a,82,124,125,136]; but see also [76]).

5. Theoretical Studies and Calculation of Nitrogen Inversion Barriers

5.1. General Considerations

One may distinguish two ways of approaching the problem of calculating nitrogen inversion barriers:

1. if one is interested mainly in barrier height, an empirical model capable of leading quickly to barrier estimates with a high enough degree of confidence may be sufficient;

2. if one is interested in the physical foundation of inversion barriers as well as in their quantitative value, it is necessary to resort to quantum chemical methods (MO calculations) which may be either of

a) semi-empirical

b) non-empirical (or *ab initio*) nature.

In case 1 the goal is to reproduce experimental phenomena, to calculate barriers from experimental data, mainly of spectroscopic and geometric origin.

In case 2 the problem is much more complex. One not only has to be concerned about the reliability of the computed barriers but one is also interested in the interpretation of the barrier heights within a given theoretical framework.

J. M. Lehn

The question is then: *why* are there barriers to inversion, i.e. what is the *physical origin* of the inversion barriers and of their dependence on structural factors? A complete answer to the question is that the total energy E_{tot} resulting from all interactions among electrons and nuclei is higher in the TS than in the GS; although this is certainly the overall barrier *origin* (!) it is however a mere tautological statement without interpretative[a] value.

The problem being of *energetic* nature, the primary, fundamental picture of the barrier origin rests on an analysis of the total energy into a complete[b] set of energy components using a certain partitioning scheme[c]. For instance E_{tot} may be separated into

V_{ne} (nuclear-electron attraction potential) $+$

V_{nn} (nuclear-nuclear repulsions) $+$

V_{ee} (bielectronic repulsions) $+$

T \quad (kinetic energy)

or into interactions between localized orbitals or bond functions.

All changes, other than energetic ones, accompanying the inversion process or affecting it (electronic redistributions, rehybridization, changes in MO features . . .) may be considered as *epiphenomena* characterizing the inversion process and reflecting the changes in total energy. The emergence and the nature of such epiphenomenological descriptions rest on the internal structure of the theoretical language used[d].

In conclusion one may distinguish three levels of methodological description

1. empirical,
2. semi-empirical,
3. non-empirical (*ab initio*),

---

[a] By "interpretative" we mean the ability of translating a gross experimental or computational *fact* (unexplanatory in itself because it is the gross result of a measurement or of the computation of a multitude of elementary interactions) into a causal relation between physical concepts and molecular structural features, the aim being to arrive at a coherent, generalized scheme applicable to present data as well as possessing predictive power.

[b] By "complete" we mean that the sum of the components is equal to the total energy.

[c] The analysis of the total energy into components corresponding to the various interactions involved carries, of course, no implication about the "nature" of energy itself [137].

[d] In this respect the "Walsh rules" [1] may be considered as epiphenomenological rules capable of predicting molecular shapes without recourse to total energy but using characteristics (MO energies) which correlate with it.

and two levels of conceptual description:

a) energetic,
b) epiphenomenological

of inversion barriers (and of energy barriers in general).

A complete a) and b) description is only possible in the non-empirical level 3; thus the specificity of *ab initio* MO calculations lies in the possibility of drawing a complete physical picture of the inversion process.

The inversion barrier is obtained as the difference in total energy between the TS and the GS and corresponds to V_{max} of Fig. 1 for an isolated molecule. Once the theoretical model is chosen, there is a further, geometrical, problem: in principle the molecular geometry should be optimized (i.e. one should search for the geometry of lowest energy) in the GS and especially in the TS. Generally the experimental geometry is taken for the GS and bond lengths and bond angles (other than angles directly affected by the inversion process) are conserved in the TS. This may be a gross oversimplification, especially for the TS, whose geometry should be optimized particularly in the case of high-quality *ab initio* calculations (see also below).

5.2. Spectroscopic Models

These models make use of spectroscopic data for the inversion vibration for calculating barrier heights.

We have already seen in Section 2 how barriers to inversion are obtained from the energy level splittings due to quantum mechanical tunneling. In this case the shape and the height of the potential curve are obtained by fitting the calculated splittings to the observed ones using Eq. (5) and (6).

A second type of model, derives the potential barrier from vibrational force constants and molecular geometrical parameters alone, assuming a certain barrier shape. Such a procedure is much less accurate (especially for high barriers) than the previous one, but it may be used for estimating barrier heights in systems for which no level splittings have been observed. It may thus be of appreciable practical usefulness.

Kincaid and Henriques [11] calculated V_{max} for a number of AX_3 molecules using a parabolic potential function suggested by Wall and Glockler [138]:

$$V = [k(|s| - |s_0|)^2]/2 \tag{19}$$

where k is the force constant for vibration v_0 [Eq. (1)],
s is the height of the nitrogen pyramid, and
s_0 is the pyramid height at equilibrium.

More recently, Costain and Sutherland [139] using a valence force model proposed the following potential for inversion vibrations in AX_3 systems:

$$V = (3/2) \, [k_1(\Delta l)^2 + k_\delta(\Delta \alpha)^2] \tag{20}$$

where k_1 and k_δ are respectively the A—X bond-streching and the X—A—X bond-bending force constants, Δl and $\Delta \alpha$ are respectively the changes in A—X bond length and X—A—X bond angle from the equilibrium values. k_1 and k_δ are obtained from the symmetrical streching ν_1 and bending ν_0 vibration frequencies respectively. Using this model Weston [6] estimated V_{max} for 12 pyramidal molecules. Koeppl et al. [140] recently used the same model for estimating V_{max} in a large number of pyramidal molecules, including cyclic systems, and extended it to the case of barriers to planar nitrogen (and carbanion) inversion.

The inversion barriers calculated in this way show that the *Costain-Sutherland model* is far superior to the earlier Kincaid-Henriques one. The agreement with experiment is satisfactory within certain limits; the results seem to be more reliable for AX_3 systems than for less symmetrical pyramidal or planar XAY systems (for an extensive study and a list of calculated barriers see [140]). Thus this spectroscopic model may provide a satisfactory estimate of barrier height in a given compound, but says nothing about the physical process itself.

5.3. Semi-Empirical Quantum Mechanical Studies

With the exception of a recent study of barriers to pyramidal inversion using the semi-empirical SCF—MO treatment MINDO [141], there are only few data on barrier calculations by semi-empirical methods.

An extended Hückel study (EHT) of *cyanamide* [143] gave a barrier of 4.5 kcal/mole and an angle φ of 22° (experimental values: 2.0 kcal/mole; 38° [33,34]). Hückel MO-type calculations on the $-O-\underset{|}{N}-O$ system taking into account the overlap integrals lead to ascribing the high barrier observed in compound *110* to increased conjugative destabilization in the transition state [72]. A very high barrier has been computed for NF_3, in agreement with expectations (Table 7; [151]).

With the exception of the MINDO results, no general conclusions can be drawn about the usefulness of the different theoretical methods in calculating nitrogen inversion barriers.

The main problem is to define the applicability and limitations of a given method.

Barriers to internal rotation have been calculated using various semi-empirical treatments; erratic results have been found in a number of

Table 7. *Calculated and experimental barriers to nitrogen inversion (in kcal/mole):*
Semi-empirical calculations

Compound	Method[a]	Ref.	Calculated Barrier	Experimental[b] Barrier (Compound)
1 NH_3	MINDO	141)	3.7	5.8 *(1)*
	INDO	151)	8.8	
	CNDO/2	152)	13.5	
5 CH_3NH_2	MINDO	141)	5.8	4.8 *(5)*
$CH_3CH_2NH_2$	MINDO	141)	5.7	—
9 $(CH_3)_2NH$	MINDO	141)	6.7	4.4 ± 1.1 *(9)*
$(CH_3)_3N$	MINDO	141)	6.5	—
13 $H_2N—CHO$	MINDO	141)	1.5	1.1 *(13)*
$H_2N—NH_2$	MINDO	141)	10.3	9.4 *(141)*; 7.4 *(142)*
16 Aziridine	MINDO	141)	13.8	$>$12 *(16)*; 18—21 *(18, 32, 43, 44)*
45	MINDO	141)	10.5	11.2 *(45)*
54 ring N	MINDO	141)	22.3	$>$22 *(54)*
54 $—NH_2$	MINDO	141)	9.7	—
11 $H_2N—CN$	EHT	143)	4.5	2.0 *(11)*
NF_3	INDO	151)	62.6	—
$H_2C=N—H$	EHT	148)	4.6	—
$H_2C=N—CH_3$	EHT	148)	13.8	25—27 *(151)*
$H_2C=N—NH_2$	EHT	148)	16.0	—
$H_2C=N—OH$	EHT	148)	23.0	$>$39 *(155)*
$H_2C=N—Br$	EHT	148)	11.0	$>$28 *(153)*
$H_2C=N—Cl$	EHT	148)	10.6	$>$31 *(154)*
$H_2C=N—F$	EHT	148)	32.5	—
$HN=NH$	MO	149)	33	$\sim$23 *(183)*; 142)c)
	EHT	150)	11.6	
	INDO	151)	46.2	
$FN=NF$	INDO	151)	67.7	32 *(182)*
	CNDO/2	153)	75.4	
$HN=C=NH$	INDO	151)	8.0	6.7 *(181)*
$HN=C=C=NH$	INDO	151)	23.9	—
$HN=C=C=C=NH$	INDO	151)	6.9	—
$FN=C=NF$	INDO	151)	22.4	—

a) MINDO: Modified INDO. INDO: Intermediate Neglect of Differential Overlap. CNDO/2: Complete Neglect of Differential Overlap, version 2. EHT: Extended Hückel Theory. MO: Molecular Orbital model (Wheland).

b) Results taken from Tables 1—6.

c) The inversion barrier is expected to be much lower (at least 5 kcal/mole) in $C_6H_5—N=N—C_6H_5$ *(183)* than in $RN=NR$ where R=H or alkyl (see also Table 6).

cases especially when heteroatoms are present [145,146]. Similar and probably even greater difficulties may be expected in studies of inversion barriers.

The MINDO calculations [141] (Table 7) reproduce satisfactorily the experimental barriers with respect to barrier heights as well as to observed trends. The barrier increase observed in hydrazines and N-chloroamines with respect to the corresponding amines has been attributed to lone pair-lone pair repulsions.

A study of the π-electronic structure of *trisilylamine* $N(SiH_3)_3$ by the Pariser-Parr-Pople SCF method indicates the presence of Si-N $(d-p)\pi$ bonding [147].

Barriers to planar nitrogen inversion have also been computed by semi-empirical methods (Table 7). EHT calculations on the $H_2C=NX$ *system* yield barriers showing the expected increase as X becomes more electroattracting ($-I$ inductive effect) [148] (expect for X = Cl, Br); the N—X overlap populations are found to decrease as the electronegativity of X and the barrier increase. The highest barriers are expected for electronegative non-conjugative substituents on nitrogen [148].

INDO calculations on the $HN=(C=)_nNH$ series ($n = 0,1,2,3$) show a barrier alternation as n increases (high barriers for $n = 0,2$; low barriers for $n = 1,3$) (Table 7; [151]).

High barriers have been computed for *N-fluorodiimide* and *N-fluoro-carbodiimide* (Table 7) in agreement with the effect expected for N—F compounds [151,153].

At the present stage it seems that the MINDO and INDO methods may produce satisfactory barriers to nitrogen inversion in a number of cases; EHT type calculations appear to be less reliable. As may be judged from the present results, the semi-empirical picture of the inversion process emerging from these studies agrees satisfactorily with the description given in Section 3.2.D. It should be noted that this picture is principally an epiphenomenological one, describing the inversion process in terms of specific electronic effects ($sp^2 \rightarrow p$ promotion, inductive effect, lone pair-lone pair repulsions, electronic population changes ...).

5.4. Non-Empirical (ab initio) Quantum Mechanical Studies

A. Computational Method and Results

Up to now all non-empirical computations of barriers to nitrogen inversion (except for ammonia) have been performed within the Hartree-Fock SCF—LCAO—MO theoretical method. Only a brief summary of the problems involved in calculating energy barriers in general and inversion barriers in particular will be given here. A more detailed discussion of the theoretical (correlation and relativistic effects) and computational (basis sets, polarization functions, geometry optimization, etc. . .) problems involved may be found elsewhere [155)e)]. The integral Hellmann-Feynmann theorem has been applied to a study of inversion barriers in *aziridine, oxaziridine* and *methylene imine* [157)].

We shall admit here that, as discussed previously [155)], correlation and relativistic effects are approximately constant in various conformations or configurations of the same molecule and that barriers to inversion may be described within the monoelectronic monoconfigurational Hartree-Fock scheme. Calculations on NH_3 [158)] and PH_3 [159)] using large basis sets (especially for NH_3) lead to optimized geometries and to an inversion barrier for NH_3 in very good agreement with the experimental values (the experimental inversion barrier of PH_3 is unknown). It thus seems that inversion barriers may be obtained quantitatively within the Hartree-Fock MO approximation. The presently available computational results lead to the following conclusions [155)]:

a) *High quality ab initio calculations* are needed if reliability and accuracy is expected; i.e. large sets of atomic basis functions should be used and polarization functions should be included (see also below).

b) *Geometry optimization* should in principle be performed both in the GS and (especially) in the TS. Although it may not be of prime importance in the case of relatively large barriers and in the GS when high quality calculations are performed (which should naturally lead to the experimental geometry), it nevertheless affects markedly the various components (see below) of the total energy (for instance in NH_3 [158)] and in PH_3 [159)]).

c) A high-quality calculation on a *model system* should be preferred to an average or mediocre calculation on actual (but larger) molecules, as it may lead to a deeper insight into the physical characteristics of the inversion process.

d) *Practically* a compromise has to be found between the size of the system, the quality of the calculation (size and composition of the basis set, geometry optimization) and the available computer time.

It is of importance to note that, using this method, inversion barriers may be computed for molecules which are too reactive or too unstable (for instance, H_2N-SiH_3, H_2N-F, oxaziridine, $CH_2=N-H$ etc. . .) or have barriers too high for experimental determinations to be performed. In addition it is possible to study the geometry (which cannot be obtained

---

e) An SCF—LCAO—MO calculation with configuration interaction of the inversion barrier in *ammonia* has been published recently. It points to the possible importance of correlation energy in determining barrier heights; however, this result is probably an artefact of the calculation (for instance, d functions on nitrogen, which are of prime importance, have not been included in the basis set) [156)].

experimentally) and the electronic structure of the TS (for instance, the N−H and P−H bond lengths in NH_3 and PH_3 respectively are shorter in the TS than in the GS) (see also [155]).

Results

Table 8 lists the barriers to nitrogen inversion obtained from non-empirical calculation; in the case of NH_3 only the result of the highest quality calculation is given (see [155] for a complete list).

We shall consider successively the energetical results and the epiphenomenological changes; in each case the relation between the *ab initio* results and the structural effects described in Section 3 will be discussed.

Effect of polarization functions. The flexibility of the basis set of atomic functions may be increased by including d type functions on the heavy atoms (especially on the inverting nitrogen) and p type functions on the hydrogen atoms. Such functions allow a polarization of the electron distribution and should not be considered as participating to bonding in the usual chemical sense ([155] and references therein).

Incorporation of d-functions on the inverting N site is especially important. Such functions contribute more to the pyramidal GS than to the TS and thus stabilize the GS with respect to the TS. In the absence of d-functions on N, the inversion barrier in NH_3 is found to be much too small and $H_2N−CN$ and $H_2N−SiH_3$ are found to be more stable in the planar form; inclusion of d-functions (and p on H) leads to the correct NH_3 barrier [158] and to pyramidal nitrogen sites in $H_2N−CN$ and $H_2N−SiH_3$ (Table 8; [144]). Of course, the relative error introduced by the absence of d-functions is smaller the higher the barrier. This may explain in part why the barriers calculated for aziridine and oxaziridine without including d-functions are in satisfactory agreement with experimental values. In PH_3, the phosphorus inversion barrier changes from 30.9 to 37.2 kcal/mole on inclusion of two sets of d-functions [159].

B. Energetic Description of Nitrogen Inversion Barriers

The total energy of a molecular state is made up of attractive and repulsive terms, the stable form of a molecule being determined by a delicate balance between them. The inversion process may be described by considering the variations of the V_{ne}, V_{nn}, V_{ee} and T components of the total energy. V_{ne} is expected to favor and V_{nn}, V_{ee} and T are expected to disfavor compact forms where the average distances among the sets of electrons and nuclei are smaller. The more compact form is expected to be the pyramidal state, unless important geometrical changes (for instance, bond length changes) occur in the TS.

Table 8. *Calculated and experimental Barriers (in kcal/mole) to nitrogen inversion: Non-empirical (ab initio) calculations*

Compound	Computational Characteristics Basis set[a]	Ref.	Calculated Barrier	Experimental Barrier (compound)[b]
1	NH_3	GTF (13.8.2/8.2) + g. opt.	158)	5.08[d]	5.8 (1)
		STO (4.2/2) with CI	156)	5.2	
9	$(CH_3)_2NH$	GTF NH (9.5.1/4 + P_z on H) CH (9.5/4) + g. opt.[c]	144)	8.6[c]	4.4 ± 1.1 (9)
11	H_2N-CN	GTF NH (9.5.1/4.1) CN (9.5)	144)	1.85[e]	2.03 (11)
13	H_2N-CHO	GTF (11.7.1/6.1) + g. opt.	160)	0.1	1.1 (13)
	H_2N-F	GTF NH (9.5.1/4.1); F (9.5)	144)	20.3	—
	H_2N-SiH_3	GTF NH (10.6.1/5.1) SiH (12.9.2/5)	144)	0.7	—
16	Aziridine ◁N–H	GTF (9.5/4)	161)	18.3[f]	18–21 (18, 32, 43, 44)
	2-Azirine ◁N–H	GTF (5.2/2)	163b)	15.5	>12 (16)
		GTF (5.2/2)	163a)	35.14	
	Oxaziridine ◁N–H	GTF (9.5/4)	161)	32.4	31–34 (80, 82–84)
	$CH_2=N-H$	GTF (10.6/5)	164,165)	26.2	25–27 (151)
	$H-N=C=N-H$	GTF (9.5/4)	165)	8.4	6.7 (181)
	$H-N=N-H$	GTF (9.5/4)	165)	50.1	—
	$H-N=N-H$ (trans)	GTF (4.2/2)	173)	72.5	—

a) GTF: basis set of Gaussian Type Functions. STO: basis set of Slater Type Orbitals. CI: Configuration Interaction. Basis set composition is given in parentheses with the following convention: (number of s functions. number of p functions on heavy atom/number of s functions. number of p functions on hydrogen atom); (s.p.d/s.p). g. opt. = geometry optimization of bond lengths and angles other than the angle φ at the inversion site. The pyramidatity angle φ or the bending angle θ (planar N sites) have been optimized in all cases.

b) Results taken from Tables 1–6.

c) TS geometry not yet optimized.

d) $\theta(<HNH)$ (calc) = 107.2°; θ(exp) = 106.7° (ref. in 158).

e) φ(calc) = 45°; φ(exp) = 38° 33,34).

f) φ(calc) = 64°; φ(exp) = 68° 12).

Total energy E_{tot} may also be partitioned into an attractive E_A and a repulsive E_R component [167)f)].

$$E_{tot} = E_A + E_R \quad \text{with} \quad E_A = V_{ne}; \; E_R = V_{nn} + V_{ee} + T \qquad (21)$$

Inversion barriers may then be qualitatively described as *attractive dominant A* or *repulsive dominant R* depending on which term E_A or E_R shows the larger positive variation (ΔA and ΔR respectively) on going from the GS to the TS [155,167)] and destabilizes the TS[g)].

Localization of the delocalized SCF—MO's leads to a description of molecular structure in terms of localized bonds and lone pairs. The total energy is then a sum of bond and lone-pair energies E_i which are made up of an "internal" (bond or lone-pair) energy component E_i^B, and a bond-bond, bond-lone pair, lone pair-lone pair, interaction component $\sum_{i \neq j} E_{ij}^{BB}$ [168)h)]:

$$E_{tot} = \sum_i E_i \qquad E_i = E_i^B + \sum_{i \neq j} E_{ij}^{BB} \qquad (22)$$

Such a description has the advantage of permitting the eventual discussion of barrier origins in terms of changes in specific local interactions.

Results:

The inversion barrier of *ammonia* is repulsive dominant; V_{ee} and V_{nn} increase more in the TS than the attractions V_{ne}, the largest variation being that of V_{ee}. Thus NH_3 is pyramidal in part because the lone-pair repels the N—H bonding electrons.

The inversion barrier in *aziridine* is attractive dominant [144,161)]; the same result has been obtained in a model system derived from NH_3 by fixing one H—N—H angle at 80° during inversion [170)]. The angle strain effect (discussed in Section 3.1) may be ascribed to the fact that the high compactness of the pyramidal form, due to the presence of a small

---

f) For a discussion of total energy partitioning see ref. [167)] and also ref. [155)].

g) It would be desirable that the A, R description of inversion barriers be not dependent on the basis set used and on geometry optimization. Although this seems to be the case for a number of rotation barriers [167)] and for the inversion barrier in NH_3 [155.158)], recent results on the barriers to rotation in H_2O_2 and H_2S_2 [169)] and on the inversion barrier in PH_3 show that changes in barrier origin may occur. In particular, optimization of the P—H bond length in the planar TS of PH_3 does not much affect the barrier height but changes the barrier origin from A (same P—H length as in the GS) to R (optimized P—H length) [159)].

h) For the theoretical definition of these terms see ref. [168)].

C—N—C or H—N—H angle, leads to a strong attractive stabilization of the GS through the V_{ne} component. Further information about the energetic nature of the angle strain effect will probably come from the analysis of the energy barrier in dimethylamine [144].

The inversion barrier in *oxaziridine* is found to be attractive dominant [161], as is the barrier in aziridine. The presence of the oxygen atom linked to the nitrogen site amplifies the variations of the energy terms during the inversion process. An analysis of the energy components indicates that there are more attractions V_{ne} in the pyramidal state and more bielectronic repulsions V_{ee} in the planar configuration of oxaziridine as compared to aziridine. Replacing a CH_2 group by an oxygen atom in aziridine leads both to an attractive stabilization of the GS and an electronic repulsive destabilization on the TS, thus increasing the inversion barrier.

However, these results do not allow a distinction to be made between the electronegativity (inductive electron withdrawal) effect and the effects due to the electronic lone-pairs on the oxygen atom.

Angle strain and electronegativity effects have been simulated in calculations on the model system H_2N—H' by artificially altering the HNH angle and increasing or decreasing the nuclear charge of H' [170]. These calculations show that:

a) increased charge on H' (increased substituent electronegativity) increases the inversion barrier and has a larger effect when the HNH angle decreases;

b) when the HNH angle is unstrained, the inversion barrier is repulsive dominant (as in NH_3) regardless of the nuclear charge on H'; when angle strain is introduced (HNH angle fixed at 80°), the barrier becomes increasingly attractive dominant as the nuclear charge on H' increases.

These results agree with those obtained on real systems (aziridine, oxaziridine, [161] and show that purely electronegative effects lead to a barrier increase, in agreement with experimental trends (Fig. 4). In addition point a) also agrees with the observed structural enhancement of substituent effects in aziridines as compared to pyrrolidines (section 3.2.C; [68]). However, the relative importance of substituent electronegativity and lone-pair effects cannot be estimated from these results. The study of the high inversion barrier computed for H_2N—F (in agreement with the effect expected for the very electronegative fluorine substituent; Table 8) should shed further light on the origin of substituent effects on inversion barriers in a real system [144].

Low barriers to inversion have been computed for *cyanamide* H_2N—CN, and *silylamine*, H_2N—SiH_3 (Table 8; [144]) in agreement with the experimental results and with expectations. The analysis of these results will allow the origin of the energetical effect of conjugative and second-row substituents to be studied. It is found that the introduction of d orbitals on silicon does not much affect the inversion barrier in silylamine [144)1)]. This result does not agree with the idea (see section 3.2.B) that nitrogen sites linked to heteroatoms bearing low-lying d orbitals should show low inversion barriers because of $(d-p)\pi$ bonding. A calculation of the inversion barrier in 2-azirine [163)] yielded a very high barrier (35 kcal/mole; Table 8) which may be attributed in part to antiaromatic destabilization of the 4π electron TS.

Methylene imine $CH_2=NH$ is a good model for studying barriers to planar nitrogen inversion. As inversion proceeds, T and V_{nn} decrease, V_{ee} is not much affected and V_{ne} increases (i.e. attractions decrease) markedly [164)]. The barrier in $CH_2=NH$ is thus attractive dominant [164, 144)], as in aziridine. In addition it is larger (26 kcal/mole) than in aziridine (18 kcal/mole). These results, together with those obtained for aziridine [161)] and H_2N-H [170)], point to the fact that the $CH_2=NH$ system may formally be considered as the *limit of maximum angle strain of a pyramidal system* (angle constrained to 0°), and as such the inversion barrier in $CH_2=NH$ should indeed be of the same type (attractive dominant) as, and higher than, the aziridine barrier1). In addition one would then also expect substituent electronegativity effects in $CH_2=N-X$ systems to be comparatively larger than in substituted aziridines. Although accurate experimental data are not available, the results listed in Table 6 *(149, 153—155)* do not disagree with this expectation.

$(p-p)\pi$ conjugation of the $N(1)$ nitrogen lone pair with the $C=N(2)\pi$ bond in *carbodiimide* $HN(1)=C=N(2)H$ lowers the inversion barrier with respect to methylene imine. An attractive dominant barrier of 8.4 kcal/mole has been computed for this compound [144,165)]. The barrier lowering with respect to methylene imine appears to arise mainly from a smaller change in the V_{ne} attractions, the variations of V_{nn}, V_{ee} and T being more similar in the two compounds.

The inversion barrier of a single nitrogen in diimide is also attractive dominant regardless of the *cis* or *trans* nature of the GS, but the changes in the individual energy components differ in both cases [144,165)].

The analysis of the inversion barriers in aziridine, oxaziridine and methylene imine in terms of localized bond (lone-pair) energies and interactions [see Eq. (22)] agrees with the above description. In addition it provides further insight into the energetic origin of the barrier. The main results of this study are as follows [168)]:

---

1) A barrier height of 0.8 kcal/mole is obtained in the absence of silicon d orbitals.

J) Such a picture is reminiscent of the Walsh model of aziridine where the nitrogen atom is sp^2 hybridized as in methylene imine [171,172)].

a) The energy of the $1s$ orbital at the inverting nitrogen site increases in the TS. $E_{1s}(N)$ destabilizes the TS mainly through bielectronic repulsions between $1s(N)$ and the bonds connected to the nitrogen[k]. Furthermore $E_{1s}(N)$ varies more in oxaziridine than in aziridine. Thus electronegative substituents increase the inversion barrier in part through changes in $1s(N)$-bond repulsions.

b) The energies of the nitrogen lone-pair n (E_n) and of the N—H bond $(E(N—H))$ decrease during inversion; thus they favour the TS with respect to the GS, contrary to what is expected on the basis of the simple picture based on $sp^x \rightarrow p$ promotion. The repulsion between n and all other bonds (in particular the N—H bond) are larger in the TS, but they compensate only in part the TS stabilization due to the E_n^B term.

c) The energy variations of the N—C and N—O bonds favour the GS. $E(N—O)$ changes much more than $E(N—C)$; this may explain the higher inversion barrier found in oxaziridine as compared to aziridine.

d) The interactions of the oxygen lone pairs (and of the C—H bonds) with the nitrogen lone pair increase from the GS to the TS.

e) The energetic changes occurring during the inversion process are not localized at the nitrogen site itself; *all* bond or lone pair energies are changed and all pairwise interactions are significant. The energetic origin of the inversion barrier is *delocalized* over the whole molecule.

It is thus not possible to attribute barrier heights and structural effects on barrier heights to an isolated or a small set of substituent factors as one would like to do (see the empirical "rules")[l]. A single local change in molecular structure not only introduces the effects of the new substituent, but also suppresses those of the former one and leads indirectly to variations of the energy terms in the remaining, supposedly "unaffected", part of the molecule.

C. Epiphenomenological Description of Nitrogen Inversion Barriers

In addition to the energetic changes described above, which are at the origin of the inversion barrier, a number of other changes, especially electronic rearrangements, occur as inversion proceeds.

Rearrangements of the electronic structure during inversion may be studied by computing the Mulliken atomic and overlap populations [174]. Atomic populations describe the electronic distribution and the charges

---

k) The change in $E_{1s}(N)$ is comparable to the barrier height; thus only calculations including $1s$ electrons lead to a correct physical description of the inversion barriers.

l) Most effects used in empirical structure-energy correlations (Section 3.2.D) are found in the localized MO analysis, which however shows that they only reflect the overall trends resulting from much more complex changes.

at atomic sites; overlap populations measure bond strenghts and attractions or repulsions between atoms.

One should not give too much quantitative meaning to such population analyses (which depend, for instance, on the atomic basis set used), but the trends in population rearrangements occurring during inversion may serve to characterize the process and compare the electronic structure of the molecular GS and TS. The following trends in electronic structure rearrangements have been found:

a) The nitrogen lone-pair MO changes from an sp hybridized orbital to a pure $2p$ orbital in the TS [161,164,144].

b) A $N(2s) \rightarrow N(2p)$ electron transfer occurs on going from the GS to the TS; the transfer is mainly from a diffuse $2s$ orbital to a less diffuse $2p$ orbital; a large transfer corresponds generally to a large inversion barrier although there is no direct correlation between the magnitude of the transfer and the barrier height from one molecule to another[m] [161, 164,165,144].

c) The bond overlap populations and the total overlap population is larger in the TS than in the GS, except for the N—O bond of oxaziridine which stays the same in the TS (in agreement with the larger increase in $E(N-O)$ found in the localized MO analysis; see above and [168]) [158,161, 144].

d) During inversion in cyanamide, electron population is transferred from the nitrogen lone pair into the cyanonitrogen p orbital, which is parallel to the lone pair orbital [144]. In carbodiimide, NH=C=NH, one finds a similar transfer from the lone pair of the inverting nitrogen into the p orbital of the second nitrogen, which is part of the C=N π bond and therefore lies in the same plane as the lone pair at the inverting site [165, 144]. Such transfers are reminiscent of the usual lone pair-π bond conjugation picture leading to low inversion barriers as found experimentally (see Section 3.2.A) and computed (Table 8).

e) In silylamine, H_2N-SiH_3, there is a small electron transfer from the nitrogen lone pair into one of the silicon d orbitals; thus weak $(d-p)\pi$ bonding seems to be present [144] although the silicon d orbitals appear not to affect much the barrier height itself (see above 5.4.B), the $(d-p)\pi$ contribution being similar in the GS and in the TS.

f) The atomic population on the inverting nitrogen has been found to increase in the TS [161,164]. However, at least in some cases, this may be an artefact due to the absence of polarization functions and to the non-optimization of TS geometry; for instance, in NH₃ the N—H bond has been found to be more polar in the GS than in the TS [158].

---

[m] Such a correlation seems to exist for O⁺, N, C⁻ inversion in the isoelectronic series $CH_2=XH$ ($X = O^+, N, C^-$) [164].

g) Walsh diagrams [1] may be constructed using the computed MO energies. They are affected by the inclusion of polarization functions. The sum of the valence electrons MO energies, $\sum \varepsilon(\text{val})$, increases from the TS to the GS as does the total energy, in agreement with Walsh's rules [1]. However the relative $\sum \varepsilon(\text{val})$ changes are not proportional to the barrier heights and the sequence in $\sum \varepsilon(\text{val})$ changes is conserved only for ammonia, aziridine and oxaziridine, but not for methylene imine [144,158,161, 164].

6. Conclusion

The present experimental studies of nitrogen inversion provide an extended but often inhomogeneous set of results, which may be rationalized using an *ad hoc* empirical and epiphenomenological picture based on a set of more or less well defined structural effects[a]. The dependence of inversion barriers on structural effects is reasonably well known and a gross estimation of the barrier height is possible in most cases, for instance for predicting the optical stability of asymmetric nitrogen sites.

The non-empirical studies of the nitrogen inversion process provide a general picture of the energetic origin of the inversion barrier. This picture, however, depends on the energy partitioning scheme used. The attractions E_A versus repulsions E_R scheme may provide an overall picture and a dichotomic A, R classification of the barriers; it does generally not allow a detailed description in terms of molecular structural features.

The E_{tot} analysis in terms of localized MO energies may provide such a description and demonstrates the complexity of the local energetic changes occurring during inversion.

It is clear that from the moment the available theoretical methods permit the correct total energy barrier to be calculated, the barrier origin *is understood* in terms of elementary interactions within the framework of quantum mechanics[b]. However, although structure-energy correlations add nothing to the basic physical understanding of the process, such correlations are strongly needed in practice as they provide a framework for rationalizing experimental or computational results using a small (but complete) set of terms.

---

[a] There is clearly a need for more accurate determinations of activation enthalpies (and entropies) in homogeneous series of compounds.

[b] For this statement to be entirely correct in the particular case of inversion barriers, it will of course be necessary to go beyond the Hartree-Fock scheme.

J. M. Lehn

Further non-empirical studies are necessary in order to establish such correlations and to devise the energy partitioning scheme(s) which may allow a general description of inversion barriers, possessing also predictive power[c].

I wish to thank Dr. B. Munsch and Dr. J. Wagner for the numerous theoretical and experimental results they made available prior to publication.

7. References

[1] Walsh, A.D.: Discussions Faraday Soc. *2*, 18 (1947); J. Chem.Soc. 2260—2330 (1953).
[2] Glasstone, S., Laidler, K. J., Eyring, H.: The theory of Rate Processes. New York: McGraw Hill Book Co. 1941.
[3] Townes, C. H., Schawlow, A. L.: Microwave Spectroscopy, Ch. 12. New York: McGraw Hill Book Co. 1955.
[4] Herzberg, G.: Infrared and Raman Spectra of Polyatomic Molecules. Princeton, N. J.: Van Nostrand Co. 1945.
[5] Dennison, D. M., Uhlenbeck, G. E.: Phys. Rev. *41*, 313 (1932).
[6] Weston, R. E.: J. Am. Chem. Soc. *76*, 2645 (1954).
[7] Berry, R. S.: J. Chem. Phys. *32*, 933 (1960).
[8] Brickmann, J., Zimmermann, H.: Ber. Bunsenges. Physik. Chem. *70*, 157, 521 (1966).
[9] — — Z. Naturforsch. *23a*, 11 (1968).
[10] — — J. Chem. Phys. *50*, 1608 (1969).
[11] Kincaid, J. F., Henriques, F. C., Jr.: J. Am. Chem. Soc. *62*, 1474 (1940).
[12] a) Kemp, M. K., Flygare, W. H.: J. Am. Chem. Soc. *90*, 6267 (1968);
b) Tolles, W. M., Gwinn, W. D.: J. Chem. Phys. *42*, 2253 (1965).
[13] Swalen, J. D., Ibers, J. A.: J. Chem. Phys. *36*, 1914 (1962).
[14] Reeves, L. W.: Advances in Physical Organic Chemistry; V. Gold, Ed. New York: Academic Press 1965.
[15] Johnson, C. S.: Advances in Magnetic Resonance; J. S. Waugh, Ed. New York: Academic Press 1965.
[16] Binsch, G.: Topics in Stereochemistry; E. L. Eliel and N. L. Allinger, Eds. New York: Interscience Publ. 1968.

---

[c] We have tried to place the particular problem of nitrogen inversion within the framework of inversion barriers and energy barriers in general. The present discussion may also apply to other inversion processes (carbanion, oxonium, phosphorus ... (for specific examples see [164,159]), each however retaining a certain specificity. In particular, one may expect very similar structural effects on inversion barriers along an isoelectronic series of inversion sites: carbanion, nitrogen, oxonium, the barriers decreasing in the order $C^- > N > O^+$. This is probably true for both pyramidal and planar inversion processes (see ref. [164]). After completion of the manuscript a recent discussion of the experimental and theoretical aspects of pyramidal inversion [175] was brought to the attention of the author.

17) a) Anet, F. A. L., Bourn, A. J. R.: J. Am. Chem. Soc. *89*, 760 (1967) and references therein;
b) Mannschreck, A., Mattheus, A., Rissmann, G.: J. Mol. Spectry *23*, 15 (1967). — Jaeschke, A., Muensch, H., Schmid, H. G., Friebolin, H., Mannschreck, A.: ibid. *31*, 14 (1969). — Rabinovitz, M., Pines, A.: J. Am. Chem. Soc. *91*, 1585 (1969). — Dahlquist, K. I., Forsen, S.: J. Phys. Chem. 4124 (1969).

18) Frost, A. A., Pearson, R. G.: Kinetics and Mechanism. New York: John Wiley and Sons 1961.

19) Davies, M.: Z. Naturforsch. *17b*, 854 (1962).

20) Grubb, E. L., Smyth, C. P.: J. Am. Chem. Soc. *83*, 4879 (1961). — Smyth, C. P.: Molecular Relaxation Processes, Chem. Soc. Spec. Publ. no. 20, Chem. Soc. and Academic Press, p. 1 (1966).

21) Garg, S. K., Smyth, C. P.: J. Chem. Phys. *46*, 373 (1967).

22) Williams, G.: Trans. Faraday Soc. *64*, 1219 (1968).

23) Allerhand, A., Gutowsky, H. S., Jonas, J., Meinzer, R. A.: J. Am. Chem. Soc. *88*, 3185 (1966).

24) Bell, R. P.: The Proton in Chemistry. Ithaca, N. Y.: Cornell University Press 1959.

25) Brot, C.: Chem. Phys. Letters *3*, 319 (1969).

26) Labhart, H.: Chem. Phys. Letters *1*, 263 (1967).

27) Menzinger, M., Wolfgang, R. L.: Angew. Chem. *12*, 446 (1969).

28) Saunders, M., Yamada, F.: J. Am. Chem. Soc. *85*, 1882 (1963).

29) Engstrom, J. P.: Massachusetts Institute of Technology Seminars in organic chemistry, p. 255. Cambridge, Mass. (first semester 1965—1966).

30) Tsuboi, M., Hirakawa, A. Y., Tamagake, K.: J. Mol. Spectry. *22*, 272 (1967).

31) Wollrab, J. E., Laurie, V. W.: J. Chem. Phys. *48*, 5058 (1968).

32) Lister, D. G., Tyler, J. K.: Chem. Commun. 152 (1966).

33) Fletcher, W. H., Brown, F. B.: J. Chem. Phys. *39*, 2478 (1963).

34) Tyler, J. K.: private communication.

35) Costain, C. C., Dowling, J. M.: J. Chem. Phys. *32*, 158 (1960).

36) Lide, D. R., Jr.: J. Chem. Phys. *38*, 456 (1963).

37) Brois, S. J.: Transactions N. Y. Acad. of Sciences Ser. II, *31*, 931 (1969).

38) Loewenstein, A., Neumer, J. F., Roberts, J. D.: J. Am. Chem. Soc. *82*, 3599 (1960).

39) Gutowsky, H. S.: Ann. N. Y. Acad. Sci. *70*, 786 (1958). — Heeschen, J. P.: Ph. D. Thesis U. of Illinois (1959).

40) Bottini, A. T., Roberts, J. D.: J. Am. Chem. Soc. *80*, 5203 (1958).

41) Brois, S. J.: J. Am. Chem. Soc. *89*, 4242 (1967).

42) Bystrov, V. F., Kostyanovsky, R. G., Panshin, O. A., Stepanyants, A. U. Yuzhakova, O. A.: Opt. Spectr. USSR (English Transl.) *19*, 122 (1965); *19*, 217 (1965) (original Russian edition).

43) Anet, F. A. L., Osyany, J. M.: J. Am. Chem. Soc. *89*, 352 (1957).

44) Boggs, G. R., Gerig, J. T.: J. Org. Chem. *34*, 1484 (1969).

45) Jautelat, M., Roberts, J. D.: J. Am. Chem. Soc. *91*, 642 (1969).

46) Lehn, J. M., Wagner, J.: unpublished results.

47) Andose, J. D., Lehn, J. M., Mislow, K., Wagner, J.: J. Am. Chem. Soc., in press.

48) Bardos, T. J., Szantay, C., Navada, C. K.: J. Am. Chem. Soc. *87*, 5796 (1965).

49) Atkinson, R. S.: Chem. Commun. 676 (1968).

50) Kostyanovsky, R. G., Samoilova, Z. E., Tchervin, I. I.: Tetrahedron Letters 3025 (1968).

51) — Tchervin, I. I., Fomichov, A. A., Samoilova, Z. E., Makarov, C. N., Zeifman, Yu. V., Dyatkin, B. L.: Tetrahedron Letters 4021 (1969).

J. M. Lehn

52) Logothetis, A. L.: J. Org. Chem. *29*, 3049 (1964).
53) Brois, S. J.: Tetrahedron Letters 5997 (1968); J. Am. Chem. Soc. *92*, 1079 (1970).
54) Anet, F. A. L., Trepka, R. D., Cram, S. J.: J. Am. Chem. Soc. *89*, 357 (1967).
55) Lehn, J. M., Wagner, J.: Chem. Commun. 148 (1968).
56) Kostyanovsky, R. G., Tchervin, I. I., Panshin, O. A.: Izv. Akad. Nauk SSSR, Ser. Khim. 1423 (1968).
57) Lehn, J. M., Wagner, J.: Chem. Commun. 1298 (1968).
58) Paulsen, H., Greve, W.: Chem. Ber. *103*, 486 (1970).
59) Kostyanovsky, R. G., Samoilova, Z. E., Tchervin: I. I.: Tetrahedron Letters 719 (1969).
60) — Chervin, I. I., Afanas'ev, A. A., Fomichev, A. A., Samoilova, Z. E.: Izv. Akad. Nauk SSSR, Ser. Khim. 726 (1969).
61) Felix, D., Eschenmoser, A.: Angew. Chem. *80*, 197 (1968).
62) Lambert, J. B., Oliver, W. L., Jr.: J. Am. Chem. Soc. *91*, 7776 (1969).
63) Mannschreck, A., Seitz, W.: Angew. Chem. *81*, 224 (1969).
64) Montanari, F., Moretti, I., Torre, G.: Chem. Commun. 1086 (1969).
65) Mannschreck, A., Linss, J., Seitz, W.: Annalen *727*, 224 (1969).
66) Fahr, E., Fischer, W., Jung, A., Sauer, L.: Tetrahedron Letters 161 (1967).
67) Lee, J., Orrell, K. G.: Trans. Faraday Soc. *61*, 2342 (1965).
68) Lehn, J. M., Wagner, J.: Tetrahedron, in press (1970).
69) Lambert, J. B., Oliver, W. L.: Tetrahedron Letters 6187 (1968).
70) Dietrich, B., Lehn, J. M., Linscheid, P.: unpublished results.
71) a) Riddell, F. G., Lehn, J. M., Wagner, J.: Chem. Commun. 1403 (1969);
    b) Lehn, J. M., Wagner, J.: unpublished results.
72) Müller, K., Eschenmoser, A.: Helv. Chim. Acta *52*, 1823 (1969).
73) Elguero, J., Marzin, C., Tizané, D.: Org. Magn. Res. *1*, 249 (1969).
74) Griffith, D. L., Olson, B. L.: Chem. Commun. 1682 (1968).
75) Raban, M., Jones, F. B., Carlson, E. H., Banucci, E., LeBel, N. A.: J. Org. Chem. *35*, 1496 (1970).
76) Anderson, J. E.: J. Am. Chem. Soc. *91*, 6374 (1969). — Anderson, J. E., Lehn, J. M.: Bull. Soc. Chim. France 2402 (1966) and unpublished results.
77) Ebsworth, E. A. V.: Chem. Commun. 530 (1966).
78) Junge, B., Staab, H. A.: Tetrahedron Letters 709 (1967).
79) Kintzinger, J. P., Lehn, J. M., Wagner, J.: Chem. Commun. 206 (1967).
80) Yousif, G. A., Roberts, J. D.: J. Am. Chem. Soc. *90*, 6428 (1968).
81) Anderson, J. E., Roberts, J. D.: J. Am. Chem. Soc. *90*, 4186 (1968).
82) — Oehlschlager, A. C.: Chem. Commun. 284 (1968).
83) — Lehn, J. M.: J. Am. Chem. Soc. *89*, 81 (1967).
84) Allred, E. L., Anderson, C. L., Miller, R. L., Johnson, A. L.: Tetrahedron Letters 525 (1967).
85) Dewar, M. J. S., Jennings, B.: J. Am. Chem. Soc. *91*, 3655 (1969).
86) Griffith, D. L., Roberts, J. D.: J. Am. Chem. Soc. *87*, 4089 (1965).
87) Rautenstrauch, V.: Chem. Commun. 1122 (1969).
88) Sudmeier, J. L., Occupati, G.: J. Am. Chem. Soc. *90*, 154 (1968).
89) Delpuech, J. J., Martinet, Y.: Chem. Commun. 478 (1968).
90) — Deschamps, M. N.: Chem. Commun. 1188 (1967). — Delpuech, J. J., Martinet, Y., Petit, B.: J. Am. Chem. Soc. *91*, 2158 (1969).
91) Buckingham, D. A., Marzilli, L. G., Sargeson, A. M.: J. Am. Chem. Soc. *89*, 3429 (1967); *91*, 5227 (1969). — Erickson, L. E., Fritz, H. L., May, R. J., Wright, D. A.: J. Am. Chem. Soc. *91*, 2513 (1969).
92) Anderson, J. E., Griffith, D. L., Roberts, J. D.: J. Am. Chem. Soc. *91*, 6371 (1969).

374

93) Steric Effects in Organic Chemistry, M. S. Newman, Ed. New York: John Wiley and Sons 1956.
94) Brauman, S. K., Hill, M. E.: J. Chem. Soc. B, 1091 (1969).
95) Eliel, E. L.: Stereochemistry of Carbon Compounds, New York: McGraw Hill Book Co. 1962.
96) Bent, H. A.: Chem. Rev. 61, 275 (1961).
97) Hendrickson, J. B.: J. Am. Chem. Soc. 83, 4537 (1961).
98) Chiang, J. F., Wilcox, C. F., Jr., Bauer, S. H.: J. Am. Chem. Soc. 90, 3142 (1968).
99) Nygaard, L., Nielsen J. T., Kirchheiner, J., Mattesen, G., Rastrup-Andersen, J., Sorensen, G. O.: J. Mol. Struct. 3, 491 (1969).
100) Larsen, I. K.: Acta Chem. Scand. 22, 843 (1968).
101) Shibaeva, R. P., Atovmyan, L. O., Kostyanosky: R. G.: Dokl. Akad. Nauk. SSSR 175, 586 (1967). — Zacharias, H. M., Trefonas, L. M.: J. Heterocyclic Chem. 5, 343 (1968).
102) Papoyan, T. Z., Chervin, I. I., Kostyanovsky, R. G.: Izv. Akad. Nauk SSSR, Ser. Khim. 1530 (1968).
103) Shagidullin, R. R., Grechkin, N. P.: Zh. Obshchei Khim. 38, 150 (1968).
104) Kostyanovsky, R. G., Prokofiev, A. K.: Izv. Akad. Nauk. SSSR, Ser. Khim. 473 (1967).
105) Robiette, A. G., Sheldrick, G. M., Sheldrick, W. S., Beagley, B., Cruickshank, D. W. J., Monaghan, J. J., Aylett, B. J., Ellis, I. A.: Chem. Commun. 909 (1968). — Glidervell, C., Rankin, D. W. H., Robiette, A. G., Sheldrick, G. M.: J. Mol. Structure 4, 215 (1969).
106) Rankin, D. W. H.: Chem. Commun. 194 (1969).
107) Hinze, J.: Fortschr. Chem. Forsch. 9, 449 (1968) (review article).
108) Gordy, W., Orville Thomas, W. J.: J. Chem. Phys. 24, 439 (1956). — Clifford, A. F.: J. Phys. Chem. 63, 1227 (1959).
109) Klopman, G.: J. Chem. Phys. 43, S 124 (1965).
110) Streitwieser, A., Jr., Holtz, D.: J. Am. Chem. Soc. 89, 692 (1967). — Streitwieser, A., Jr., Marchand, A. P., Pudjaatmaka, A. H.: J. Am. Chem. Soc. 89, 693 (1967).
111) Gillespie, R. J.: J. Chem. Educ. 40, 295 (1963).
112) Raban, M., Kenney, G. W. J., Jr.: Tetrahedron Letters 1295 (1969).
113) — — Jones, F. B., Jr.: J. Am. Chem. Soc. 91 6677 (1969).
114) Imbery, D., Friebolin, H.: Z. Naturforsch. 23d, 759 (1968). — Cowley, A. H., Dewar, M. J. S., Jackson, W. R.: J. Am. Chem. Soc. 90, 4185 (1968).
115) Carreira, L. A., Lord, R. C.: J. Chem. Phys. 51, 2735 (1969).
116) Gerig, J. T.: J. Am. Chem. Soc. 90, 1065 (1968). — Jensen, F. D., Beck, B. H.: J. Am. Chem. Soc. 90, 1066 (1968).
117) Glazer, E. S., Knorr, R.: unpublished work reported in: Roberts, J. D.: Chem. Brit. 529 (1966).
118) Anet, F. A. L., Hartman, J. S.: J. Am. Chem. Soc. 85, 1204 (1963).
119) St. Jacques, M., Brown, M. A., Anet, F. A. L.: Tetrahedron Letters 5947 (1966).
120) Wurmb-Gerlich, D., Vögtle, F., Mannschreck, A., Staab, H. A.: Ann. Chem. 708, 36 (1967).
121) Curtin, D. Y., Hausser, J. W.: J. Am. Chem. Soc. 83, 3474 (1961).
122) Rieker, A., Kessler, H.: Tetrahedron 23, 3723 (1967).
123) Vögtle, F., Mannschreck, A., Staab, H. A.: Annalen 708, 51 (1967).
124) Kessler, H., Leibfritz, D.: Tetrahedron 25, 5127 (1969).
125) — — private communication.
126) — Tetrahedron Letters 2041 (1968).

127) Marullo, N. P., Wagener, E. H.: J. Am. Chem. Soc. *88*, 5034 (1966); Tetrahedron Letters 2555 (1969).
128) McCarthy, C. G., Wieland, D. M.: Tetrahedron Letters 1787 (1969).
129) Bauer, V. J., Fulmor, W., Morton, G. O., Safir, S. R.: J. Am. Chem. Soc. *90*, 6847 (1968).
130) Ogden, P. H., Tiers, G. V. D.: Chem. Commun. 527 (1967).
131) Anet, F. A. L., Jochims, J. C., Bradley, C. H.: J. Am. Chem. Soc. *92*, 2557 (1970).
132) Binenboym, J., Burcat, A., Lifshitz, A., Shamir, J.: J. Am. Chem. Soc. *88*, 5039 (1966).
133) Bürgi, H. B., Dunitz, J. D.: Chem. Commun. 472 (1969).
134) Wettermark, G., Weinstein, J., Souza, J., Dogliatti, L.: J. Phys. Chem. *69*, 1584 (1965).
135) Andreades, S.: J. Org. Chem. *27*, 4163 (1962).
136) Lehn, J. M., Price, B. J., Riddell, F. G., Sutherland, I. O.: J. Chem. Soc. *B*, 387 (1967).
137) Theobald, D. W.: The concept of Energy, London: E. and F. N. Spon Ltd. 1966.
138) Wall, F. T., Glockler, G.: J. Chem. Phys. *5*, 314 (1937).
139) Costain, C. C., Sutherland, G. B. B. M.: J. Phys. Chem. *56*, 321 (1952).
140) Koeppl, G. W., Sagatys, D. S., Krishnamurthy, G. S., Miller, S. I.: J. Am. Chem. Soc. *89*, 3396 (1967) and revision of some results by the same authors.
141) Dewar, M. J. S., Shanshal, M.: J. Am. Chem. Soc. *91*, 3654 (1969).
142) Hartley, G. S.: J. Chem. Soc. 633 (1938). — Halpern, Brady, Winkler: Can. J. Res. *B 28*, 140 (1950).
143) Henneike, H. F., Drago, R. S.: J. Am. Chem. Soc. *90*, 5112 (1968).
144) Lehn, J. M., Munsch, B.: unpublished results and work in progress.
145) Herndon, W. C., Feuer, J.: Tetrahedron Letters 2625 (1968).
146) De La Vega, J. R., Fang, Y., Hayes, E. F.: Intern. J. Quant. Chem. *3S*, 113 (1969).
147) Perkins, R. G.: Chem. Commun. 268 (1967).
148) Kerek, F., Simon, Z., Ostrogovich, G.: J. Chem. Soc. in press (1970).
149) Wheland, G. W., Chen, P. S. K.: J. Chem. Phys. *24*, 67 (1956).
150) Alster, J., Burnelle, L. A.: J. Am. Chem. Soc. *89*, 1261 (1967).
151) Gordon, M. S., Fischer, H.: J. Am. Chem. Soc. *90*, 2471 (1968).
152) Lewis, T. P.: Tetrahedron *25*, 4117 (1969).
153) Herndon, W. C., Feuer, J., Hall, L. H.: Theoret. Chim. Acta *11*, 178 (1968).
154) Kitzing, R., Fuchs, R., Joyeux, M., Prinzbach, H.: Helv. Chim. Acta *51*, 888 (1968).
155) Lehn, J. M.: Theoretical Conformational Analysis. *Ab initio* SCF—LCAO—MO studies of conformations and conformational energy barriers. Scope and limitations. Proceedings of the International Symposium on Conformational Analysis, Brussels, 8—12 September 1969; Academic Press, to be published.
156) Pipano, A., Gilman, R. R., Bender, C. F., Shavitt, I.: Chem. Phys. Letters *4*, 583 (1970).
157) Vinh-Leveille, J.: Thèse de Doctorat de 3ᵉ Cycle, Université de Paris (1970).
158) Rauk, A., Allen, L. C., Clementi, E.: J. Chem. Phys. *52*, 4133 (1970).
159) Lehn, J. M., Munsch, B.: Chem. Commun. 1327 (1969) and unpublished results.
160) Bak, B.: private communication.
161) Lehn, J. M., Munsch, B., Millié, Ph., Veillard, A.: Theoret. Chim. Acta *13*, 313 (1969).
162) Lehn, J. M., Wagner, J.: Chem. Commun. 414 (1970).

[163] a) Clark, D. T.: International Symposium, Quantum Aspects of Heteroatomic Compounds in Chemistry and Biochemistry, Jerusalem (1969); to be published by the Israel Academy of Sciences and Humanities. See also Chem. Commun. 637 (1969);
b) Clark, D. T.: Chem. Commun. 850 (1969).

[164] Lehn, J. M., Munsch, B., Millié, Ph.: Theoret. Chim. Acta, *16*, 351 (1970).

[165] — — Theoret. Chim. Acta *12*, 91 (1968).

[166] Dewar, M. J. S., Jennings, B.: Tetrahedron Letters 339 (1970).

[167] Allen, L. C.: Chem. Phys. Letters *2*, 597 (1968).

[168] Lévy, B., Millié, Ph., Lehn, J. M., Munsch, B.: Theoret. Chim. Acta, in press (1970).

[169] Veillard, A.: Theoret. Chim. Acta, in press (1970); Veillard, A., Demuynck, J.: Chem. Phys. Letters *4*, 476 (1970).

[170] Rauk, A., Allen, L. C., Mislow, K.: in preparation.

[171] Walsh, A. D.: Trans. Faraday Soc. *45*, 179 (1949).

[172] Kochanski, E., Lehn, J. M.: Theoret. Chim. Acta *14*, 281 (1969).

[173] Schaad, L. J., Kinser, H. B.: J. Phys. Chem. *73*, 1901 (1969).

[174] Mulliken, R. S.: J. Chem. Phys. *23*, 1833, 1841, 2338, 2343 (1955).

[175] Rauk, A., Allen, L. C., Mislow, K.: Angew. Chem., in press.

[176] Gribble, G. W., Easton, N. R., Jr., Eaton, E. T.: Tetrahedron Letters 1075 (1970).

[177] Dahm, J., Fehér, F.: Z. Naturforsch., *B*, in preparation. — Dahm, J.: Diplomarbeit, Chemistry Department, University of Köln (1970).

[178] Kessler, H., Leibfritz, D.: Tetrahedron Letters 1423 (1970).

[179] Lehn, J. M., Munsch, B.: Chem. Commun., to be published.

[180] Kessler, H.: Angew. Chem. *82*, 237 (1970).

[181] Lambert, J. B.: Topics in Stereochemistry, Interscience Publ., to be published.

Received March 24, 1970

Note added into proof

The inversion barriers in fluoramine, in cyanamide and in silylamine are found to be *attractive* dominant [144], in contrast to the repulsive barriers one might expect on the basis of the calculations on the H_2N-H' model system where only "electronegativity" effects are considered (see page 367).

Recent references:

Nitrogen inversion in compounds of type *138 B* (N—CH_3 derivative; $\Delta G_c^{\neq} \sim 14$ kcal/mole) [176].

Nitrogen inversion and hindered rotation processes in Aziridyl-S_n-Aziridyl ($n = 1,2, 3,4$) compounds [177].

Nitrogen inversion in imines [178].

ab initio computations on H_2N-SiH_2 [179].

Reviews on inversion processes [180,181].

Konformative Beweglichkeit von Siebenring-Systemen

Doz. Dr. W. Tochtermann

Institut für Organische Chemie der Universität Heidelberg

Inhalt

Conformational Mobility of Seven-Membered Ring Systems

Abstract

This article provides a general review of the conformational mobility of seven-membered ring systems. Most of the results discussed were obtained between 1964 and 1969.

After a short qualitative introduction to the principles of "dynamic nuclear magnetic resonance spectroscopy", the proposed interconversion processes for cycloheptanes and cycloheptenes are explained in detail. According to calculations, the most favourable conformation for cycloheptanes seems to be the twist chair; cycloheptenes prefer the chair form. Possible conformational processes for chair and boat forms are discussed and illustrated.

Next, other unsaturated carbocyclic systems are discussed, such as dibenzo-cycloheptadienes (atropisomeric biphenyls), dihydropleiadenes, cycloheptatrienes and their benzo-derivatives. Most of these compounds exist in a half-boat or boat conformation. Tropones are considered in the light of recently published work.

Structural data so far obtained from X-ray analyses, electron diffractions, microwave spectra, NMR-coupling constants etc. are also mentioned.

In these series, the energy barriers for the ring inversion of the seven-membered ring may sometimes become very high (E > 20—30 kcal/Mol), so that relatively stable conformational isomers or enantiomers can be obtained. In such cases the corresponding energy parameters can be determined from classical equilibration studies or from racemisation kinetics. Moreover, conformational rigid molecules (which have been found in the series of bridged biphenyls, dihydropleiadenes and substituted tribenzocycloheptatrienes) are useful for studies of the relationship between conformation and chemical reactivity. Remarkable differences in the relative reactivity of diastereotopic hydrogens and of conformational isomers have been found some seven-membered ring systems.

Chapter 3 summarizes the results for nitrogen, oxygen and sulfur heterocycles. In some examples the heteroatom was varied, to enable its influence on conformations and their mobility to be discussed.

An especially interesting situation is found in the field of heterocycles containing more than one sulfur atom. Here two different relatively stable conformations (f.i. chair and twist-boat form) have often been detected in the NMR-spectra; they are interconverted by two processes (version and pseudorotation) which differ significantly in the height of their energy barrier. In a few cases the mobile heterocycles crystallize in only one of these conformations.

Finally, there is a brief mention of the hypothesis that the conformation of the seven-membered ring in certain compounds may influence their biological activity.

Einleitung

Die rasche Entwicklung der Konformationsanalyse [1, 2, 2a)] in den letzten Jahrzehnten dürfte vornehmlich zwei Ursachen haben: Einmal konnte D. H. R. Barton 1950 [3)] die Beziehungen zwischen Konformation und chemischer Reaktivität — am Beispiel axialer und äquatorialer Substituenten in Cyclohexan-Derivaten — aufzeigen und damit das Interesse der an Reaktionen interessierten Chemiker auf Konformationsprobleme lenken.

Zum anderen wurden durch die Anwendung physikalischer Untersuchungsmethoden zunehmend Aussagen über die Konformation zahlreicher Verbindungen sowie über Geschwindigkeit und Aktivierungsparameter von konformativen Vorgängen, z. B.

Rotationen um Bindungen,
Umklappvorgänge bei Ring-Systemen,
Inversions- und Pseudorotationsphänomene an zahlreichen Atomen

zugänglich. Hierbei sind die *Röntgenstrukturanalyse* (für den kristallisierten Zustand) die *Elektronenbeugung* (für die Gasphase), die *UV-, IR- und Mikrowellen-Spektroskopie*, die optische *Rotationsdispersion* und der *Zirkulardichroismus*, die Anwendung von *Ultraschallmethoden*, die Untersuchung der optischen Aktivität und vor allem die dynamische *Kernresonanzspektroskopie* zu nennen [4]. Der zuletzt genannten Methode kommt besondere Bedeutung zu, da die Temperaturabhängigkeit geeigneter Signale in den Kernresonanz-Spektren häufig Einblick in die intramolekulare Beweglichkeit von Molekülen gibt.

Es wird versucht, die erhaltenen Ergebnisse an carbo- und heterocyclischen Siebenringsystemen zusammenzufassen. Dabei werden vor allem Fragen der Konformation und der inneren Beweglichkeit besprochen; außerdem wird auf konformativ bedingte Reaktivitätsunterschiede sowie auf das Problem „Konformation und pharmakologische Wirkung" kurz eingegangen.

1. Untersuchung intramolekularer Vorgänge mit der NMR-Spektroskopie [5-8]

Die theoretischen Grundlagen [9,11,12], die methodischen und experimentellen Verfahren [14] sowie die Aussagemöglichkeiten [10,13,14] der Kernresonanzspektren organischer Verbindungen sind bekannt. Auch über NMR-spektroskopische Untersuchungen von Umklappvorgängen bei Ringsystemen („Ringinversionen") liegen Zusammenfassungen vor [13,14]. Da der weitaus größte Teil der hier zu besprechenden Resultate mit dieser Methode erhalten wurde, sei sie an Hand eines einfachen Beispiels qualitativ erläutert.

Die Kernresonanzspektroskopie ist zum Studium schneller und reversibler Vorgänge unter Gleichgewichtsbedingungen vorzüglich geeignet, da das NMR-Spektrum oft durch die Lebensdauer einer Konformation entscheidend geprägt wird [5-14]. Das zu betrachtende System befindet sich dabei im Zustand des thermodynamischen Gleichgewichts zwischen verschiedenen oder identischen Spezies — etwa A und A' —; makroskopische Veränderungen können also nicht beobachtet werden.

Dieser Fall liegt etwa im 9.9-Dimethoxy-1.4-diaza-9 H-tribenzo-[a.c.e]-cyclohepten *(1)* vor, dessen ^{1}H-NMR-Spektrum in Hexachlor-butadien/1.1.2.2-Tetrachloräthan (1:1) bei 33 °C zwei getrennte scharfe Singuletts für die axialen (a) und äquatorialen (e)[a] Methoxylgruppen bei $\tau_a = 7{,}38$ und $\tau_e = 6{,}50$ zeigt [15]. Aus dieser Nichtäquivalenz folgt

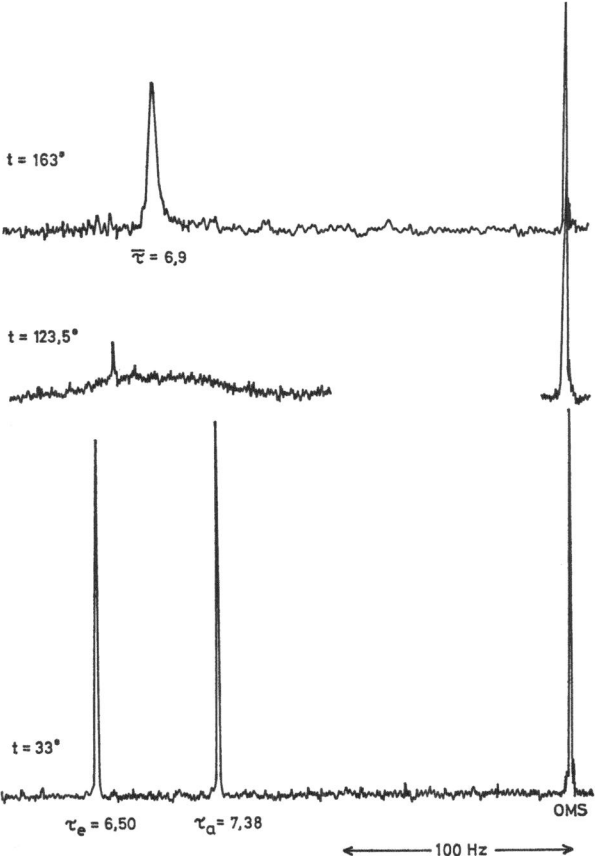

Abb. 1. Signale der Methoxylprotonen von 9.9-Dimethoxy-1.4-diaza-9 H-tribenzo-[a.c.e]-cyclohepten *(1)* in C_4Cl_6/$CHCl_2CHCl_2$ (1:1) bei verschiedenen Temperaturen (OMS = Octamethylcyclotetrasiloxan)

---

[a] Obwohl die Terminologie ,,axial'' und ,,äquatorial'' streng nur für das Cyclo-hexan-System definiert wurde, soll sie hier aus Gründen der Einfachheit auch für Siebenringe verwendet werden. Die den axialen bzw. äquatorialen Lagen im Cyclohexan vergleichbaren Positionen bei Siebenring-Systemen werden in der Literatur häufig auch als pseudo- oder quasi-axial bzw. pseudo- oder quasi-äquatorial bezeichnet.

einmal, daß *1* in einer nicht-ebenen Boot- oder Wannen-Konformation vorliegt [16]. Bei Erhöhung der Temperatur verbreitern sich nun diese Signale, um bei $T_c = 123{,}5\ °C$ zu einer einzigen, noch breiten Absorption zusammenzufließen. T_c wird allgemein als Koaleszenz- oder *Aufspaltungstemperatur* bezeichnet. Bei noch höheren Temperaturen wird diese Resonanz nun immer schärfer (d. h. die Halbwertsbreite der Resonanzlinien wird immer kleiner), bis sie schließlich bei etwa 160 °C in ein einziges scharfes Signal für alle sechs Methoxylprotonen übergeht.

Demnach muß also in *1* ein Austauschprozeß zwischen axialen und äquatorialen Substituenten vor sich gehen, der bewirkt, daß bei höherer Temperatur nur eine mittlere Absorption bei $\bar{\tau} = 6{,}9$ erhalten wird. Am Modell ist leicht erkennbar, daß die ursprüngliche a-OCH$_3$-Gruppe ($\rightarrow$ e'-OCH$_3$-Gruppe) sich nach Umklappen des zentralen Siebenringes in *1* $\left(A \underset{k_I}{\overset{k_I}{\rightleftharpoons}} A' \right)$ nachher in der chemischen Umgebung der e-OCH$_3$-Gruppe befindet und umgekehrt, d. h. diese beiden Reste vertauschen beim Ringinversionsvorgang ihre Positionen relativ zum Ring.

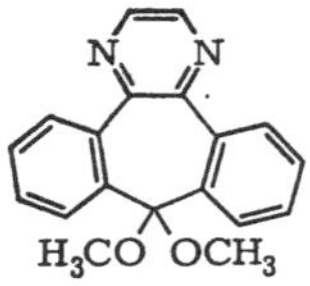

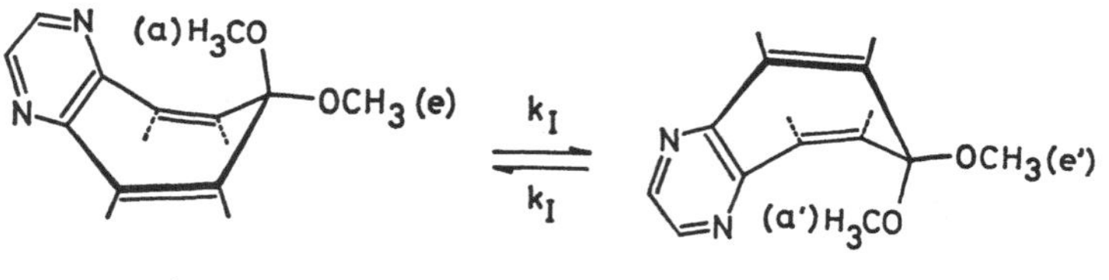

Nun hängt es von der Geschwindigkeitskonstanten k_I der Ringinversion ab, ob für die beiden diastereotopen [16a] axialen und äquatorialen OCH$_3$-Gruppen eine Aufspaltung in zwei Linien mit entsprechenden chemischen Verschiebungen τ_a und τ_e oder nur eine einzige Resonanz bei $\bar{\tau}$ zwischen τ_a und τ_e erhalten wird.

Wie am Beispiel von *1* illustriert wurde, können nun durch Variation der Temperatur oft die Extremfälle — kleines k_I, große Lebensdauer t bzw. großes k_I, kleine Lebensdauer t — verwirklicht und die jeweilige Gestalt der Signale untersucht werden. Da somit die Signalform von der Geschwindigkeit des Austauschprozesses (bei *1* k_I) abhängt, können hieraus durch Verwendung exakter Gleichungen oder von Näherungen

Geschwindigkeitskonstanten in einem breiten Temperaturbereich erhalten werden. Hieraus ergeben sich dann unter Verwendung der Eyring- und Arrhenius-Beziehungen die Aktivierungsparameter für den betrachteten Prozeß ($\Delta G_T^{\neq}, \Delta H_T^{\neq}, E_A, \Delta S_T^{\neq}, \log k_0$) [17,18]. So gilt für den Fall der Verbindung *1* bei der erwähnten Aufspaltungstemperatur T_c unter gewissen, bei *1* erfüllten Voraussetzungen die einfache Beziehung

$$k_{\mathrm{I}}(T_c) = \frac{\pi \cdot \Delta \nu}{\sqrt{2}} = \frac{\pi \cdot 53}{\sqrt{2}} = 120 \ \mathrm{sec}^{-1} \ (\text{bei } T_c = 123,5\,^\circ\mathrm{C}).$$

($\Delta \nu$ = Differenz der Chemischen Verschiebungen für die (a)- und (e)-OCH$_3$-Gruppen (in Hz) bei langsamem Umklappen), so daß die Freie Enthalpie der Aktivierung ($\Delta G^{\neq} = R \cdot T_c \ln k_{\mathrm{B}} \cdot T_c/h \cdot k_{\mathrm{I}} = 19,7$ kcal/Mol) [18] bei T_c für diesen Prozeß besonders leicht erhältlich ist ($k_{\mathrm{B}} =$ Boltzmann-Konstante, $h =$ Plancksches Wirkungsquantum). Bei Gleichgewichten mit ungleicher Besetzung von A und A' sind natürlich über die relativen Intensitäten der Signale auch die entsprechenden Gleichgewichtsparameter ($\Delta G_T, \Delta H; \Delta S_T$) aus dem Spektrum zugänglich.

Insgesamt lassen sich so intramolekulare Vorgänge untersuchen, deren Energieschwellen etwa zwischen 5—6 kcal/Mol als unterer und 20—25 kcal/Mol als oberer Grenze liegen. Da zahlreiche intramolekulare Bewegungen in diesen Bereich fallen, hat sich diese Methode [19] als außerordentlich fruchtbar erwiesen.

Leider sind auch Fehlermöglichkeiten des Verfahrens der Linienverbreiterung bekannt [14,20], so daß vor der unkritischen Anwendung von Näherungsbeziehungen gewarnt werden muß [14]. Die Ermittlung der Geschwindigkeitskonstanten durch „complete line shape analysis" wird daher als unerläßlich angesehen [14,158].

Ein von Schmid, Friebolin, Kabuß und Mecke ausgearbeitetes Parameter-Verfahren [21], das es erlaubt, über leicht meßbare Größen die Geschwindigkeitskonstanten aus Diagrammen zu entnehmen, ist nach ausführlichen kinetischen Untersuchungen [22] dem complete line shape analysis-Verfahren ungefähr gleichwertig. Diese Diagramme wurden gleichfalls unter Verwendung eines Computers aus den exakten Gleichungen [23] gewonnen. Mannschreck [22,24] konnte an Benzamiden und Nitrosaminen (Rotationsvorgänge um die C—N- bzw. N—N-Bindung) zeigen, daß die aus Linienverbreiterungen mit dem Parameter-Verfahren erhaltenen kinetischen Daten mit denjenigen übereinstimmen, die aus klassischen Äquilibrierungen gewonnen werden [24a]. Damit war die dynamische Kernresonanz-Spektroskopie [156,157] auf eine sichere experimentelle Grundlage gestellt.

2. Carbocyclische Verbindungen

2.1. Cycloheptan-, Cyclohepten- und Monobenzo-cycloheptenderivate [13,14]

Berechnungen von Hendrickson [25,26] ergaben für Cycloheptan zwei bevorzugte Konformationen, die Twist-Sesselform *2* (TS) und die etwas ungünstigere Twist-Wannen- oder Twist-Boot-Form *3* (TW) mit C_2-Symmetrie. Im Gegensatz zur Sesselform des Cyclohexans sind hier jedoch beide Formen flexibel; der Übergang von Substituenten zwischen axialen und äquatorialen Positionen in *2* kann durch Pseudorotation [27] ohne merkliche Winkeldeformationen sehr schnell erfolgen.

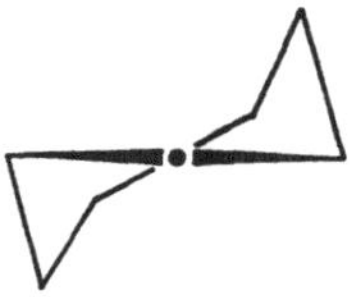

2 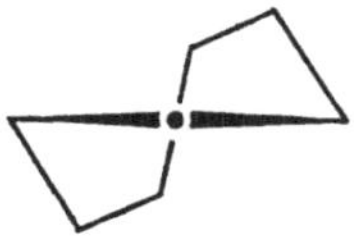   *3*

Dieser Prozeß benötigt im allgem. sehr wenig Energie und kann daher mit der Methode der Linienverbreiterung oft nicht mehr erfaßt werden. (Bei Schwefel-Heterocyclen wurden jedoch stark erhöhte Pseudorotationsbarrieren gefunden; s. S. 430.)

^{19}F-NMR-Untersuchungen von Roberts [28] am *1.1-Difluorcycloheptan 4* ergaben, daß bis herab zu $-180\,°C$ keine Nichtäquivalenz der beiden Fluoratome beobachtet werden konnte. Dies kann einmal dahingehend gedeutet werden, daß selbst dann noch alle intramolekularen Umklappvorgänge sehr schnell verlaufen; andererseits ist es auch möglich, daß die Fluoratome sich in der 1-Position der Twist-Sessel-Form *4* ($\triangleq$ *2*) befinden. Dann sind aus Symmetriegründen die Fluorsubstituenten äquivalent und es kann keine unterschiedliche chemische Verschiebung beobachtet werden.

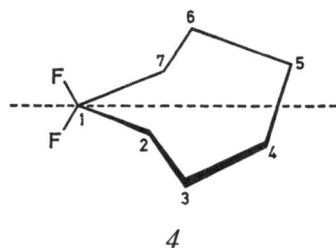

4

Für alle anderen Stellungen — außer am C^1 — ergeben sich jedoch verschiedene (äquatoriale bzw. axiale) Lagen für Substituenten. Beim Cycloheptan gibt es zwei Möglichkeiten, wie nicht-äquivalente Substituenten ihre Positionen vertauschen können: Neben einem Umklapp-vorgang, der dem beim Cyclohexan [1,2,3] ähnelt und bei dem eine Twist-Sessel- (oder Sessel-) Form in eine Twist-Boot- (oder Boot-) Form und dann wieder in eine neue Sessel-Form übergehen würde („Version"), ist eben auch ein Pseudorotationsvorgang mit einer

(Twist-Sessel)-(Sessel)-(Twist-Sessel)-Form-Umwandlung

zu diskutieren [25]. Der Austausch äquatorialer und axialer Reste würde immer dann erfolgen, wenn ein substituiertes Kohlenstoffatom die mit 1- bezeichnete Position der Twist-Sessel-Form durchläuft.

Nach Roberts [28] können die zum Umklappen eines monosubstituierten Cycloheptans notwendigen Pseudorotationsschritte, bei denen R sozusagen um den Ring herumwandert, für die Sesselformen folgendermaßen dargestellt werden:

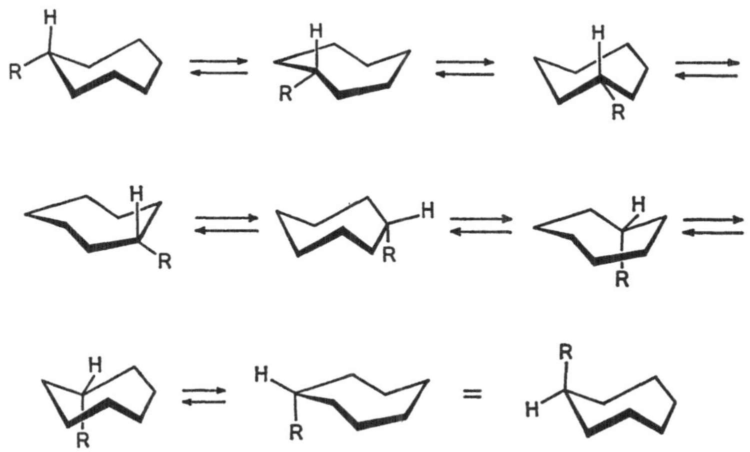

Die Pseudorotation des *Cycloheptans* erfordert nach Hendrickson [25] nur eine Aktivierungsenergie von etwa 2 kcal/Mol.

Grundsätzlich kann aber die Umwandlung zweier Sessel- (oder Twist-Sessel-) Konformationen ineinander zwei Vorgänge umfassen [29, 141]:

1. einen Pseudorotationsvorgang
2. einen Versionsvorgang, womit $(S \rightleftharpoons W)$-Prozesse gekennzeichnet werden sollen, die unter vorübergehenden Valenzwinkeldeformationen erfolgen.

Bei Schwefel-Heterocyclen ist es gelungen, zwischen den verschiedenen Möglichkeiten zu unterscheiden. [29,141] (s. S. 430).

Zusammenfassend ergeben sich folgende Möglichkeiten für die gegenseitige Überführung der symmetrieausgezeichneten Konformationen des gesättigten Siebenringes [141]

Sessel (S)- und Wannen (W)-Form mit C_s (C_{1v})-Symmetrie sowie Twistsessel (TS)- und Twistwannen (TW)-Form mit C_2-Symmetrie

1. die Pseudorotationen der *Sesselreihe*, d. h. die Schritte

$S \rightleftharpoons TS \rightleftharpoons S \rightleftharpoons TS$ usw.

Bei jedem Pseudorotationsvorgang wandert das Symmetrieelement um eine halbe Bindungslänge, wobei sich die Symmetrie der Molekel ändert

$$S \longrightarrow TS \quad \text{bedingt} \quad C_s \longrightarrow C_2.$$

Nach insgesamt 14 gleichsinnig erfolgenden Pseudorotationsschritten ist das Symmetrieelement 7 Ringatome gewandert und befindet sich wieder in der ursprünglichen Lage; das Molekül besitzt dann die gleiche Symmetrie wie am Anfang aber alle ehemals axialen Liganden stehen äquatorial und umgekehrt, d. h. es hat *Ringinversion* stattgefunden. Auf S. 385 sind die sieben zu durchlaufenden Sesselformen für den Fall des Umklappens einer S-Konformation wiedergegeben; zwischen je zweien liegt eine (nicht eingezeichnete) TS-Konformation. Nach weiteren 14 Pseudorotationsschritten wäre dann die Ausgangskonformation wieder erreicht.

Ein ähnlicher Pseudorotationscyclus kann für die *Wannenreihe* (W $\rightleftharpoons$ TW usw.) aufgestellt werden.

2. die Versionsvorgänge mit Übergängen zwischen den beiden Reihen (S $\rightleftharpoons$ W, S $\rightleftharpoons$ TW, TS $\rightleftharpoons$ W, TS $\rightleftharpoons$ TW).

Die Ringinversion kann demnach auf zwei verschiedenen Wegen ablaufen (s. S. 384/385), einmal durch Pseudorotation innerhalb einer Reihe oder aber durch eine Folge von Versions- und Pseudorotationsschritten, z. B.

$$TS \rightleftharpoons TW \rightleftharpoons W \rightleftharpoons \quad \rightleftharpoons W \rightleftharpoons TW \rightleftharpoons TS.$$

Während beim Cycloheptan das Umklappen des Siebenringes ausschließlich durch Pseudorotation erfolgen sollte [25], sind für substituierte und heterocyclische Derivate oft beide Möglichkeiten zu berücksichtigen, da dort einzelne Pseudorotationsschritte stark erschwert sein können.

Versuche am *1.1.3.3-Tetrafluorcycloheptan 5* [28] zur Unterscheidung der Umklappmechanismen waren ergebnislos. Bei dieser Verbindung sollte zumindest ein Paar von geminalen Fluorsubstituenten nicht-äquivalent sein, wenn die Pseudorotation und/oder die Ringinversion eingefroren wäre; beide Prozesse verlaufen daher in *5* sehr rasch.

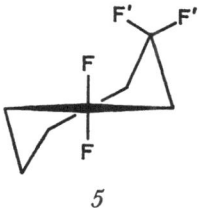

5

Dagegen zeigte *1.1-Difluor-4.4-dimethyl-cycloheptan 6* [28] im ^{19}F-NMR-Spektrum bei tiefen Temperaturen ($< -152\,°C$) zwei verschiedene Fluorresonanzen. Hier besetzen vermutlich die größeren geminalen Methylgruppen die 1-Position*, was zwangsläufig zu nicht identischen Fluorsubstituenten führt, wenn die intramolekularen Bewegungen langsam werden.

Die Auswertung der Linienverbreiterungen von *6* ergibt eine Aktivierungsenergie von etwa 6 kcal/Mol; da wegen der geminalen Methylgruppen für die Pseudorotation eine höhere Energieschwelle (9,6 kcal/Mol [30]) zu erwarten ist, wird angenommen, daß es sich bei dem einfrierbaren Prozeß eher um eine Ringinversion, an der Versionsschritte beteiligt sind, handelt [28].

Dem Pseudorotationsprozeß wird hier deshalb eine höhere Energieschwelle zugeschrieben, da Konformationen mit nach innen stehender Methylgruppe, die beim ,,Herumwandern dieser Substituenten um den Ring'' durchlaufen werden müssen, sterisch ungünstig sind [28,30].

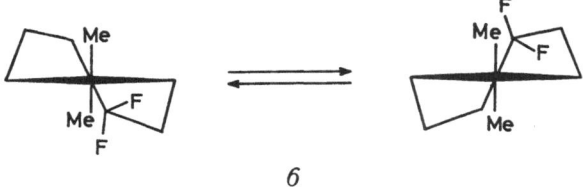

6

Ein interessanter Fall stellt *4.5-trans-Dibrom-1.1-difluor-cycloheptan 7* [31] dar, das im Protonen-entkoppelten ^{19}F-NMR-Spektrum oberhalb von $-113\,°C$ ein einziges Fluorsignal zeigt, welches unterhalb von $-118\,°C$ in zwei Singuletts im Intensitätsverhältnis 74:26 aufspaltet. Diese beiden Resonanzen können am besten wieder den Twist-Sessel-Konformationen *7a* und *7b* zugeschrieben werden, in denen die Fluoratome die Positionen am C^1 besetzen, somit innerhalb jeder Konforma-

---

*) Bei unsymmetrischer Substitution liegt keine Symmetrieachse mehr vor.

tion a oder b aus Symmetriegründen identisch sind und daher auch keine F—F-Kopplung zeigen. Die stark positive Aktivierungsentropie von $\Delta S^{\pm} = +15 \pm 3$ e.u. dieses Vorgangs ($\Delta H^{\pm} = 9{,}8 \pm 0{,}3$ kcal/Mol) ist besser mit der Annahme einer Umwandlungsfolge

flexible Twist-Sesselform $7a$ $\rightleftharpoons$ (flexible Boot- bzw. Twist-Boot-form) $\rightleftharpoons$ flexible Twist-Sesselform $7b$

zu vereinbaren als mit einem Pseudorotationsvorgang, bei dem ähnlich wie bei 6 sterisch ungünstige Konformationen durchlaufen werden müßten.

Für die Cycloheptan-Reihe liegen außerdem Konformationsanalysen mehrerer Derivate [31a)] sowie eingehende Untersuchungen von B. Eistert [31b)] über den Zusammenhang zwischen Enolisierungstendenz und Beweglichkeit des Cycloheptandions-(1.3) vor.

Für den ungesättigten Siebenring im Cyclohepten 8 ($R^1 = R^2 = H$) ergeben Modellbetrachtungen [29)] eine starre Sesselform $8a$ und eine flexible Bootform (Wannenform), wobei deren energieärmste Konformation nach Hendrickson [25)] nicht das normale Boot $8b$, sondern das Twist-Boot $8c$ darstellt.

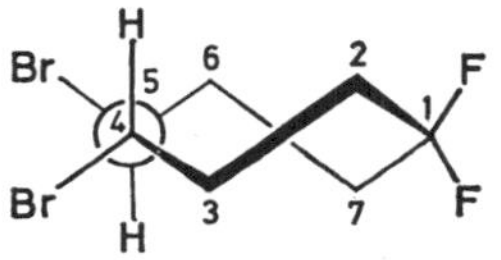

7a (74%) 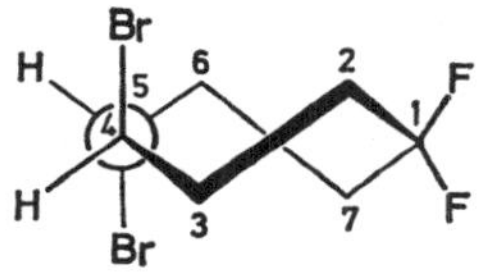   7b (26%)

Nach Friebolin, Mecke, Kabuß und Lüttringhaus [32)] können folgende Kriterien zur NMR-Konformationsanalyse derartiger ungesättigter Ringsysteme herangezogen werden:

Die Inversion einer Sesselform in eine zweite energiegleiche ist nur über eine Wanne unter Valenzwinkeldeformation möglich, beinhaltet demnach einen Versionsvorgang. Die beiden enantiomeren „Twist-Formen" (C_2-Symmetrie) lassen sich dagegen durch Pseudorotation über die normale Wanne hinweg ineinander überführen. Für den ersten Vorgang ist die relativ höhere, für den zweiten die relativ niedrigere Energiebarriere zu erwarten [b)]. In der Sessel-Form $8a$ unterscheiden sich Substi-

---

b) Allerdings zeigen die auf S. 392 besprochenen Modellrechnungen, daß sich die Versions- und Pseudorotationsbarrieren bei Cycloheptenen erheblich weniger unterscheiden als bei Cyclohexanen [35)]. Der Schluß, daß z.B. ein Übergang aufgrund der Höhe einer gefundenen Energiebarriere ein Versionsvorgang sein muß, ist daher nicht gerechtfertigt. Für eine eindeutige Zuordnung sind zusätzliche Informationen notwendig (vgl. S. 430–432).

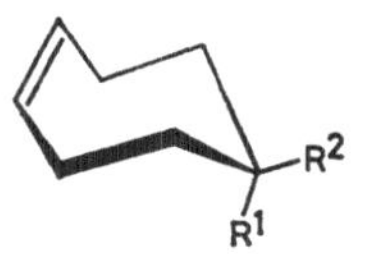

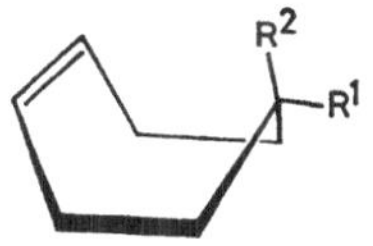

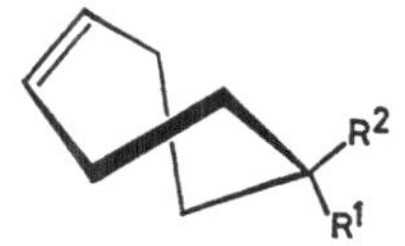

Symmetrie: C_s C_s C_2
            S   W   T

8a   *8b*   *8c*

tuenten am C^5 durch eine äquatoriale bzw. axiale Lage, während in der Twist-Bootform *8c* durch C^5 wiederum die zweizählige Symmetrieachse verläuft. Beim Einfrieren der Umklappvorgänge sollten also R^1 und R^2 in *8a* unterschiedliche, in *8c* identische Signale aufweisen. Ist etwa $R^1 = R^2 = H$, so würde es sich in *8a* um diastereotope, in *8c* dagegen um äquivalente Protonen handeln [4b,16a].

Da *5.5-Difluor-cyclohepten* [31] (8, $R^1 = R^2 = F$) bei tiefen Temperaturen im protonenentkoppelten ^{19}F-NMR-Spektrum für die beiden Fluorsubstituenten ein AB-System zeigt, das bei höheren Temperaturen ($T_c = -92\,°C$) zusammenfließt, darf der Schluß gezogen werden, daß diese Verbindung in der Sesselkonformation *8a* vorliegt. Die Auswertung der Spektren zwischen $-147\,°C$ und $-44\,°C$ führt zu den Aktivierungsparametern $\Delta H^{\neq} = 7,4 \pm 0,1$ kcal/Mol und $\Delta S^{\neq} = 0,2 \pm 1$ e.u. für die Ringinversion, durch die R^1 und R^2 ausgetauscht werden.

Ebenso ergeben die ^1H-NMR-Spektren der Benzocyclohepten-Derivate *9*, *10* und *11* bei tiefen Temperaturen AB-Systeme für die Methylenprotonen bzw. nicht-äquivalente Methylgruppen in den bezifferten Stellungen. Diese Signale zeigen zwischen $-57\,°C$ (bei *9*) und $-45\,°C$ (bei *10* und *11*) Koaleszenz, woraus sich die Freien Enthalpien der Aktivierung zu $\Delta G^{\neq}_{T_c} = 11-12$ kcal/Mol für die Ringinversion der Sesselform (entsprechend *8a*) bestimmen lassen [33]. Bei *9* wurde E_A zu $13,0 \pm 1,5$ kcal/Mol ermittelt. Im Tieftemperatur-Spektrum von *9* ist der A-und B-Teil des Quartetts der Methylenprotonen am C^5 von gleicher Intensität. Da außerdem keine zusätzlichen Signale beobachtet werden, dürfte im Benzo-cyclohepten mindestens zu 95% ein einziges Konformeres vorliegen.

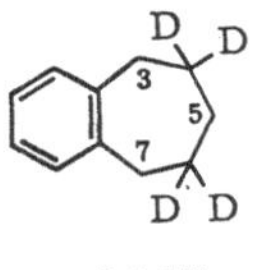

9 [155)] 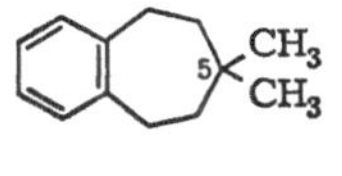   *10* 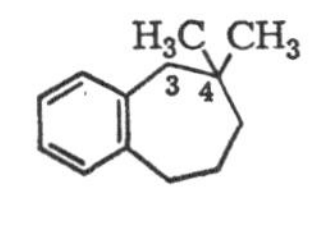   *11*

Diese Untersuchungen wurden auf mehr als zwanzig weitere Benzo-cyclohepten-Derivate ausgedehnt [34,35]. Die Resultate lauten:

Bei *9, 10, 12* und *13* erscheinen in den Tieftemperatur-Spektren („eingefrorene" Konformationen!) die Protonenresonanz-Signale der diastereotopen Liganden am C^3 des Ringes bei denselben chemischen Frequenzen wie die der entsprechenden Liganden am C^7; das gleiche gilt für Substituenten am C^4 und C^6. Diese Moleküle müssen also ein Symmetrieelement besitzen, das die genannten Ringpositionen äquivalent oder zumindest enantiotop macht [16a]. Dabei kann es sich um eine zwei-zählige Achse oder eine Symmetrieebene handeln. Dieser Befund zeigt, daß auch bei Benzocycloheptenen nur die oben genannten Konforma-tionen *8a—8c* (S, W oder T) für den ungesättigten Siebenring zu berück-sichtigen sind. Wegen der aufgefundenen Nichtäquivalenz der geminalen Substituenten am C^5 (diastereotope Substituenten!) kann auch die Twist-Boot-Form T mit C_2-Symmetrie ausgeschlossen werden.

Die Protonen der C^3/C^7- und C^4/C^6-Methylengruppen ergeben jeweils bei tiefen Temperaturen nur ein AB-Quartett, geminale Methylgruppen unter diesen Bedingungen jeweils nur ein Dublett. Bei diesen Verbin-dungen ist also ein einziges Konformeres energetisch ganz besonders begünstigt $(c \geqq 95\%)$.

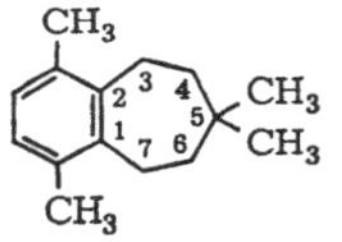

12 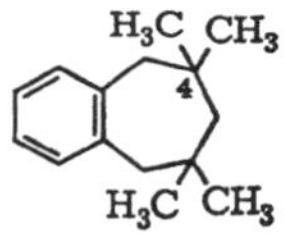   *13*

Eine Unterscheidung zugunsten der Sessel- und gegen die Wannen-form kann aufgrund folgender Argumente getroffen werden: In der Wannenform würde ein axialer Substituent am C^5 aufgrund seiner Lage oberhalb der Ebene des 1.2-ständigen Benzolringes gegenüber dem äqua-torialen eine beträchtliche chemische Verschiebung (ca. 1—2 ppm) nach höherer Feldstärke erfahren. Die gemessenen, weitaus geringeren Diffe-renzen $(\varDelta\delta \approx 0,2\,\text{ppm})$ sind nach Modellrechnungen nur mit der Sesselform vereinbar. Die gleiche Schlußfolgerung ergab sich aus der Analyse der vicinalen Kopplungskonstanten zwischen Protonen an C^3 (bzw. C^7) einerseits und C^4 (bzw. C^6) andererseits [34,35].

Modellrechnungen führen gleichfalls zur Sesselform als energieärmster Konformation für *9, 10* und *13*; bei *9* und *13* ist die Wanne (W), bei *10* dagegen die Twist-Konformation (T) rechnerisch als zweitgünstigste

Konformation ermittelt worden[c]. Nach den [1]H−NMR-spektroskopischen Ergebnissen beträgt für *9* und *10* die Differenz der Freien Konformationsenthalpien ΔG zwischen Sessel- und der zweitstabilsten Konformation mehr als 1,8 kcal/Mol. Nach Modellbetrachtungen [35] können nun die konformativen Umwandlungen der Cyclohepten-Sessel auf verschiedenen Wegen erfolgen, die in Abb. 2 zusammengefaßt sind:

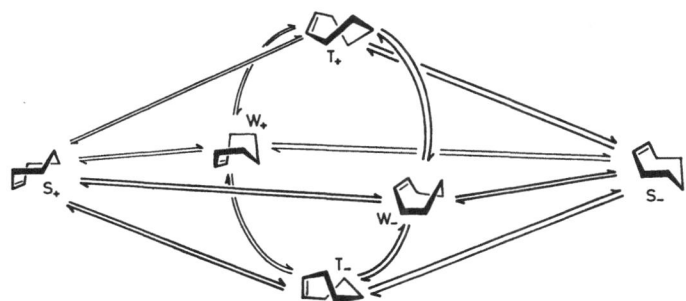

Abb. 2. Schema der konformativen Umwandlungsmöglichkeiten bei Cycloheptenen. Hierbei symbolisieren die von S_+ und S_- radial ausgehenden Pfeile Versionsumwandlungen, die gekrümmten, zu einem Kreis geschlossenen Pfeile dagegen die Pseudorotationsschritte [35]

Da die direkte Umwandlung der beiden Sesselformen S_+ und S_- über einen planaren Übergangszustand ausgeschlossen werden kann, ist die Sessel-Inversion (wie beim Cyclohexan) nur auf dem Wege über die Wannen- oder Twist-Form möglich. Dabei sind folgende Wege denkbar:

1. eine Folge zweier Versionsvorgänge, z.B. $S_+ \rightleftharpoons T_- \rightleftharpoons S_-$;

2. eine Aufeinanderfolge von Versions- und Pseudorotationsschritten,

$$z. B. \quad S_+ \rightleftharpoons W_+ \rightleftharpoons T_+ \quad oder$$
$$T_- \rightleftharpoons W_- \rightleftharpoons S_-.$$

Durch die Energieschwellen der einzelnen Teilschritte wird nun der Ablauf der Ringinversion ($S_+ \rightleftharpoons S_-$) bestimmt.

Modellrechnungen für Benzocyclohepten führten zu dem Ergebnis, daß der energetisch am wenigsten kostspielige Weg ($E_{ber} = 14{,}6$ kcal/Mol) von dem Sessel zur Wanne und zwar von S_+ nach W_+ bzw. S_- nach

---

[c] Sowohl bei der Wannen- als auch bei der Twist-Form soll es sich um Konformationen handeln, die durch Potentialminima im Energieprofil ausgezeichnet sind ([4a], S. 156).

W_ führt, wobei die Übergangskonformationen (Energiemaxima) dadurch gekennzeichnet sind, daß die zwei benachbarten Dieder-Winkel ω_{45} und ω_{56} an den Bindungen C^4—C^5 und C^5—C^6 annähernd null sind [35].

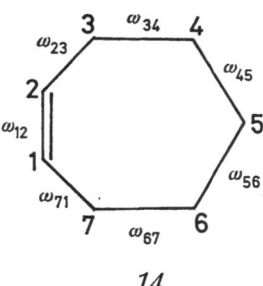

14

Andere Umwandlungen, wie $S \rightarrow T$ und $S_+ \rightarrow W_-$, bei denen die Dieder-Winkel ω_{34} oder ω_{67} bzw. ω_{23} und ω_{71} als „Umwandlungskoordinaten" verwendet wurden, erfordern einen weitaus höheren Energieaufwand ($E \approx 20$—25 kcal/Mol).

Die *Aktivierungsenergie* der Pseudorotation ($W \rightarrow T$) beim Cyclohepten-Ring wurde rechnerisch zu **11,1** kcal/Mol ermittelt, ein Wert, der erheblich über demjenigen des Cyclohexans liegt. Untersuchungen an ungesättigten siebengliedrigen Heterocyclen (s. S. 430) zeigen in der Tat, daß die Aktivierungsenergie dieses Vorganges bei Siebenringen höher als bei Sechsringen ist.

Der geschwindigkeitsbestimmende Teilschritt der Sessel-Sessel-Inversion ist demnach die Umwandlung

$$S_+ \rightarrow W_+ \text{ bzw. } S_- \rightarrow W_-;$$

die für *9* experimentell zu $13 \pm 1,5$ kcal/Mol bestimmte Aktivierungsenergie [33] der Ringinversion stimmt gut mit dem theoretisch berechneten Wert von 14,6 kcal/Mol überein.

Neben der ausführlichen kinetischen Untersuchung der Ringinversion des Tetradeuterio-cycloheptens *9* wurden noch die Freien Enthalpien der Aktivierung $\Delta G^{\ddagger}_{T_c}$ für die Sessel-Inversion bei insgesamt zwanzig Benzocycloheptenen bestimmt. Dabei ergaben sich — in Abhängigkeit von der Zahl und Stellung der Substituenten — Werte zwischen 9,9 und 13,7 kcal/Mol [35].

Ein interessanter Fall liegt in dem von Grunwald und Price [36] studierten *3.3.6.6-Tetramethyl-1.2-benzocyclohepten-1 (15)* vor.

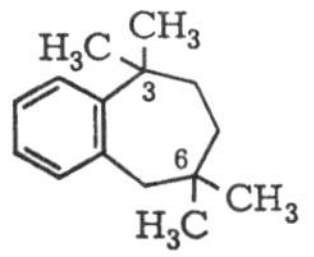

H₃C CH₃

15

Dort beobachtet man bei −81 °C im ^{1}H−NMR-Spektrum für die zum Benzolring β-ständigen Methylgruppen (Stellung 6) vier verschiedene Resonanzen, von denen jeweils zwei von gleicher Intensität sind. Dieser Befund lehrt, daß hier zwei verschieden stabile Konformere mit jeweils zwei diastereotopen Methylgruppen vorliegen. Dabei handelt es sich wohl um die Sessel- und Wannenform, die im Verhältnis 2:1 ($\Delta H = 0{,}3$ kcal/Mol) auftreten. Bemerkenswerterweise ergeben die Modellrechnungen der Freiburger Gruppe [34] für 15 eine ähnliche Stabilität von Sessel- und Wannenform ($\Delta E \approx 0$), womit der geringe, experimentell ermittelte Unterschied von $\Delta H = 0{,}3$ kcal/Mol [36] in guter Übereinstimmung steht. Demnach sind die Wannen- oder Twist-Formen bei Cycloheptenen relativ günstiger als bei Cyclohexanen.

Bei Erhöhung der Temperatur verändern sich die Spektren von 15 in einer Weise, die sich nur mit dem Auftreten zweier verschiedener dynamischer Prozesse deuten läßt. Der mit einer niedrigeren Energiebarriere $\Delta H_1^{\ddagger} = 9{,}3$ kcal/Mol) ablaufende Vorgang wird von diesen Autoren einer Sessel-Wannen (Boot)-Umwandlung, der energetisch ungünstigere ($\Delta H_2^{\ddagger} = 12{,}6$ kcal/Mol) einer Sessel-Sessel-Umwandlung ($S_+ \rightarrow S_-$; Ringinversion) zugeschrieben [36]. Nach Auffassung von Kabuß, Schmid, Friebolin und Faißt [35] ist diese Zuordnung jedoch nicht eindeutig, da es sich bei einem dieser Prozesse auch um eine Pseudorotationsumwandlung ($W_+ \rightleftharpoons T \rightleftharpoons W_-$) handeln könnte.

2.2. Dibenzocycloheptadiene (6.7-Dihydro-5 H-dibenzo[a.c]-cycloheptene, 10.11-Dihydro-5 H-dibenzo[a.d]cycloheptene) und Dihydropleiadene

Ein Dibenzo-cycloheptadien-System (Dibenzo[a.c]cyclohepten-System) liegt bei den einfach und doppelt verbrückten Biphenylen der allgemeinen Struktur 16 und 17 (X = CR₂) vor. Diese Verbindungsklasse beanspruchte im Zusammenhang mit der Atropisomerie der Biphenyle [4a,4b,]

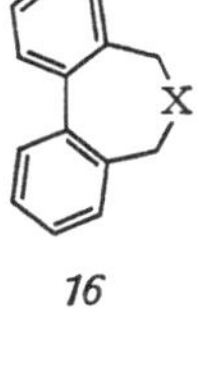

16 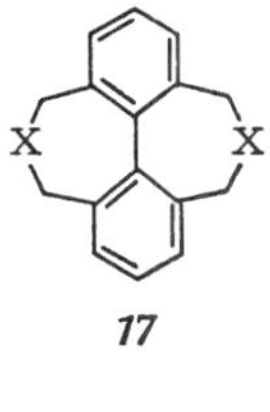   **17**

[37,38)] besonderes Interesse und wurde vor allem von Mislow u. Mitarb. eingehend studiert.

Da jedoch in dieser Reihe auch zahlreiche Heterocyclen (X = Heteroatom) untersucht wurden, folgt die zusammenfassende Diskussion aller Resultate erst im Abschnitt 3.3.3.

Nach Modellbetrachtungen [38,39)] nimmt der Siebenring in diesen Verbindungen eine verdrillte (Boot)-Konformation *18* ein und kann über einen Übergangszustand *19*, der sich durch eine koplanare Anordnung des Biphenyl-Systems auszeichnet (Torsionswinkel um die Biphenyl-1.1'-Bindung $\phi = 0$), in das Enantiomere *20* übergehen.

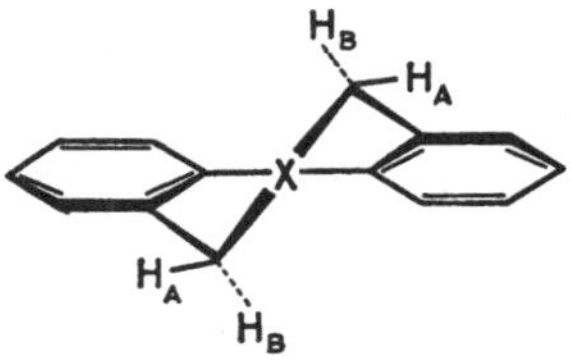

18 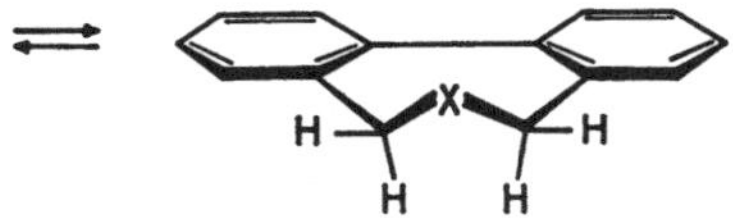   *19*

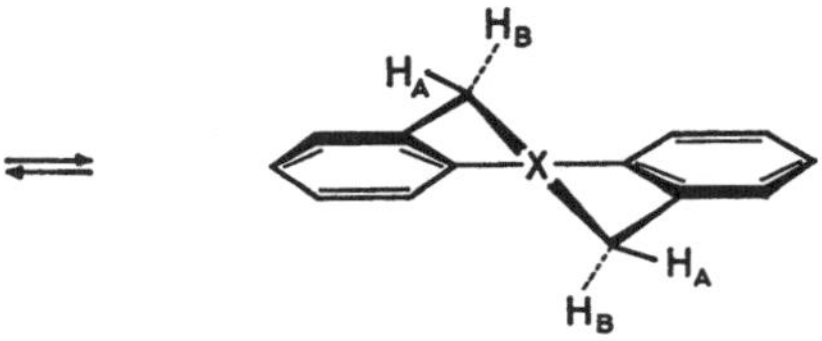

20

Verläuft dieser Vorgang genügend langsam, so können im ¹H-NMR-Spektrum die Protonen H_A und H_B der Methylengruppe als AB-System erscheinen, welches bei Temperaturerhöhung Koaleszenz zeigt [38,39)]. Dieser Fall liegt z. B. bei den von Sutherland und Ramsay [39)] untersuchten Verbindungen *21* und *22* vor, für deren Umklappvorgang (*18* → *20*) Freie Enthalpien der Aktivierung von $\Delta G^{\ddagger}_{10\,°C} = 13,7$ bzw. $\Delta G^{\ddagger}_{9\,°C} = 14,0$ kcal/Mol ermittelt wurden. Die Umwandlung der enantiomeren Konfor-

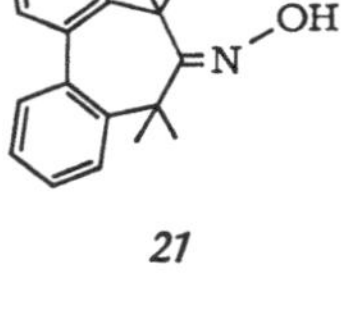

21 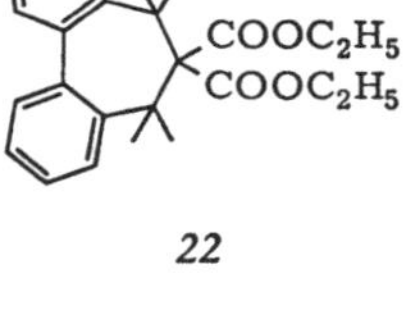   *22*

mationen erfolgt demnach bei *22* so rasch, daß die berichtete Synthese [40] des optisch aktiven Diesters *22* wohl einer Überprüfung bedarf [39].

Aus den Beziehungen zwischen Aktivierungsparametern und Geschwindigkeitskonstanten läßt sich ableiten [18,14], daß die Energiebarrieren für konformative Umwandlungen von Enantiomeren mindestens etwa 20 kcal/Mol betragen müssen, damit ein polarimetrischer Nachweis optisch aktiver Konformationsisomerer bei Raumtemperatur möglich wird.

Dieser Wert ($E_A = 20,8$ kcal/Mol) wird z.B. gerade bei dem chiralen Dibenzocycloheptadien (Dibenzo[a.d]cyclohepten) *23* erreicht, dessen Hydrogen-L-malat bei 20 °C mit einer Halbwertszeit von 12 min mutarotiert [42]. Die Erhöhung der Energieschwelle und die somit mögliche Trennung diastereomerer Salze ist hier auf eine zusätzliche Behinderung der intramolekularen Bewegungen durch die Substituenten an der semicyclischen Doppelbindung zurückzuführen (vgl. Abschnitt 2.4.1).

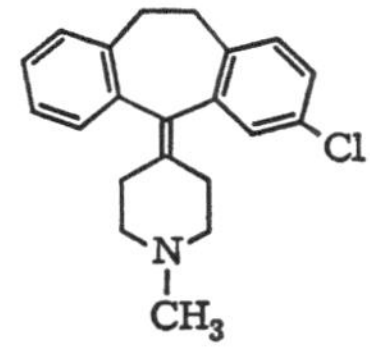

23

Während der einfach verbrückte siebengliedrige Kohlenwasserstoff *16* ($X = CH_2$) noch sehr beweglich ist (E_A(ber) ≈ 13 kcal/Mol) [38], können zahlreiche carbo- und auch heterocyclische (vgl. S. 420 ff.) Vertreter der allgemeinen Struktur *24* mit größeren Resten in den o.o'-Positionen des Biphenyl-Systems bei Raumtemperatur optisch aktiv erhalten werden und zeigen große konformative Stabilität [38,43,130].

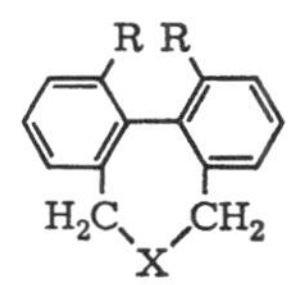

24

R = NO_2; X = CO, CHOH
R = NO_2; X = C(COOC_2H_5)_2
R = NH_2; X = C(COOC_2H_5)_2
R = CH_3; X = C(COOC_2H_5)_2

Das gleiche gilt für analoge Binaphthyle *(25)*.

Aus dem ¹H-NMR-Spektrum des Trimethylbenzo-cycloheptadiens *(26)* folgt, daß dieser Kohlenwasserstoff ausschließlich in der Konformation *A* vorliegt, da die alternative Anordnung *B* der Methylgruppen sehr ungünstig ist [39].

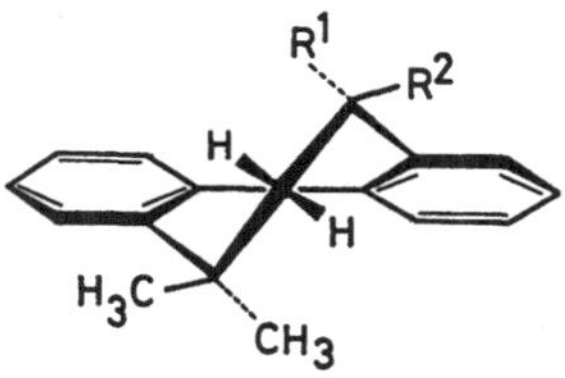

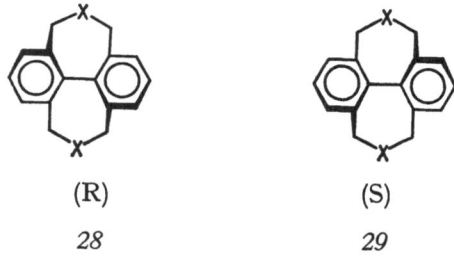

26

A: $R^1 = H$; $R^2 = CH_3$
B: $R^1 = CH_3$; $R^2 = H$

Von den doppelt verbrückten Biphenylen mit carbocyclischem Siebenring wurden die Diketone *17* (X = CO; CH₂ α zu CO) und sein Octadeutero-Derivat *27* (*17*, X = CO; CD₂ statt CH₂) untersucht; diese Verbindungen gehören zu den ersten chiralen Molekülen der Punktgruppe D₂ [4b], die optisch aktiv erhalten werden konnten [38] (weitere Beispiele s. S. 426, 433).

Am Beispiel von *17* und *27* wurde auch erstmals ein *sekundärer kinetischer H/D-Isotopieeffekt* für eine reine konformative Umwandlung gefunden: Genaue polarimetrische Messungen bei 95° zeigen nämlich, daß (−)−*17* 1.06 mal schneller racemisiert als (−)−*27* [38].

Sowohl durch die Stereochemie der kinetischen Racematspaltung [37] (partielle asymmetrische Reduktion von (±)−*17* mit Aluminium-(+)−2-octanolat) als auch durch die ORD-Kurve (negatives Vorzeichen des Carbonyl $n \rightarrow \pi^*$-Cotton-Effektes um 305 mμ) [43] kann (−)−*17* die Chiralität einer (S)-Konfiguration *29* zugeordnet werden. [38]

(R)   (S)

28   *29*

Die ORD- [43] und CD- [44] -Kurven zahlreicher optisch aktiver Biaryle (z. B. auch des Typs *24, 25* und *17*), deren absolute Konfiguration [37] zum Teil schon bekannt war, sind eingehend untersucht worden. Die daraus abgeleiteten Zusammenhänge zwischen dem Vorzeichen des Cotton-Effektes und der absoluten Konfiguration erlauben die Aufstellung von Regeln zur Bestimmung der letzteren aus den entsprechenden Kurven [38,43,44].

Die heterocyclischen Analoga zu *17* und *24* werden in Abschnitt 3 besprochen.

Die Stereochemie von *7.12-Dihydro-pleiadenen (30)* ist vor allem von Lansbury u. Mitarb. untersucht worden; eine neuere Zusammenfassung der Ergebnisse liegt vor [45].

Dieses nicht-ebene System, in dem der zentrale Siebenring eine Art Boot-Konformation („half-boat-conformation") [45] einnimmt, ist deshalb reizvoll, weil je nach Relation und Stellung der Substituenten verschiedene stereochemische Beziehungen zwischen den durch konformative Umwandlungen ineinander übergehenden Spezies bestehen können [45,46]:

1. Ist $R^1 = R^2$, so gehen — auch im Fall $R^1 = R^2 \neq H$ — bei trans-Stellung von R^1 und R^2 die Konformeren beim Umklappen des Siebenringes A $\rightleftharpoons$ B in sich selbst über, d. h. hier ist $A \equiv B$ (Beispiele *30a—30e*).

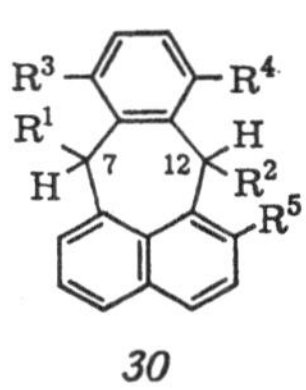

30

2. In den Fällen *30f—30g* — sinngemäß bei den Methylen-dihydropleiadenen *31a* und *31b* — bedeutet der Vorgang A $\rightleftharpoons$ B die Umwandlung von Enantiomeren ineinander.

3. Ist dagegen $R^1 \neq R^2$, so sind *A* und *B* Diastereomere (Beispiele *33*).

Verbindungen der ersten Gruppe, für die $R^1 = R^2 \neq H$ gilt, könnten nicht durch Ringinversion racemisieren, wenn sie einmal in optisch aktiver Form vorlägen. Im Gegensatz dazu würden *30f* und *30g* sowie *31a* und *31b* ihre optische Aktivität nur dann beibehalten, wenn die Energiebarriere zwischen *A* und *B* höher als 20 kcal/Mol wäre. Optisch aktive Derivate der letzten Gruppe ($R^1 \neq R^2$) könnten dagegen als isolierbare Diastereomere vorliegen und je nach der Höhe der Schwelle zwischen *A* und *B* Mutarotation eingehen.

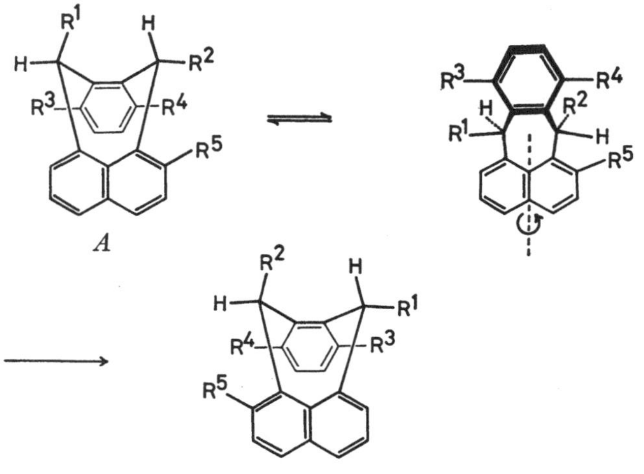

A

B

30a: $R^1=R^2=R^3=R^4=R^5=H$
30b: $R^1=R^2=H$; $R^3=R^4=CH_3$; $R^5=H$
30c: $R^1=R^2=OH$; $R^3=R^4=R^5=H$
30d: $R^1=R^2=OAc$; $R^3=R^4=R^5=H$
30e: $R^1=R^2=OCH_3$; $R^3=R^4=R^5=H$
30f: $R^1=R^2=R^4=R^5=H$; $R^3=CH_3$
30g: $R^1=R^2=R^3=R^4=H$; $R^5=OCH_3$

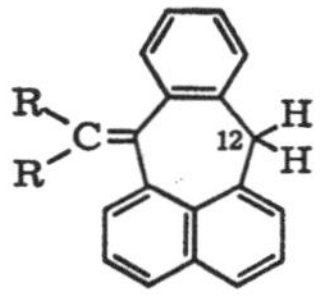

31

31a: $R=H$ 31b: $R=CH_3$

Die ¹H-NMR-Spektren von *30a*, *30b*, *30f* und *31a* ergeben für die Methylengruppen bei tiefen Temperaturen AB-Systeme, die zwischen $T_c = +8°$ (bei *30a*) und $+45$ °C (bei *31a*) Koaleszenz zeigen und bei weiterem Erwärmen in ein einziges Signal übergehen; die daraus zugänglichen Freien Enthalpien der Aktivierung für die Konformations-Inversion ($\Delta G_{T_c}^{\neq}$) liegen zwischen 13,6 und 15,6 kcal/Mol. Ganz analog konnte bei *30d* und *30e* die Temperaturabhängigkeit der Methylprotonen der

Acetoxy- und Methoxygruppen zur Ermittlung der Umklappschwellen herangezogen werden ($\Delta G^{+}_{+7°} = 14{,}3$ kcal/Mol bei *30d*; $\Delta G^{+}_{+23°} = 15{,}2$ kcal/Mol bei *30e*) [45,46,47]. Demnach ist diese Verbindungsklasse noch viel zu beweglich, um etwa bei Raumtemperatur die Isolierung von Konformations-Enantiomeren oder Isomeren in geeigneten Fällen (Gruppe 2 oder 3) zu erlauben. Eine Ausnahme in dieser Reihe machte das Isopropyliden-Derivat *31b*, so daß dort entsprechende Spaltungen aussichtsreich sind. Unter Verwendung des in der 12-Position axial monodeuterierten Kohlenwasserstoffs *31b* konnte durch Äquilibrierung ΔG^{+} zu 31 kcal/Mol bei 130 °C bestimmt werden [45]. In Einklang damit führt die partielle Dehydratisierung von 7-Isopropyl-7-hydroxy-7.12-dihydropleiaden mit (+)-Camphersulfonsäure zu linksdrehendem *31b* [46].

Die *Erhöhung der Energiebarrieren* von $\Delta G^{+} = 13{,}6$ kcal/Mol bei *30a* nach 14,5 bei *30f* und 15,6 kcal/Mol bei *30b* spricht nach Ansicht von Lansbury u. Mitarb. [46] dafür, daß sich C^7 und C^{12} in *A* beim Umklappen gleichzeitig nach unten bewegen und ein planarer Übergangszustand durchlaufen wird. Bei einem chiralen Übergangszustand für *30f* sollte der entsprechende ΔG^{+}-Wert näher bei demjenigen von *30a* liegen; es würde dann nämlich derjenige Weg bevorzugt werden, der eine möglichst geringe Kompression von C^7 und der C^8-CH_3-Gruppe zur Folge hat [46].

Die Vergleiche von Freien Enthalpien der Aktivierung, die sich streng genommen auf verschiedene Temperaturen beziehen, sind hier sicher erlaubt, da die Bestimmung sämtlicher Aktivierungsparameter mit Hilfe des „Complete line-shape-analysis-Verfahrens" [48] bei einigen Dihydropleiadenen zu von Null nur wenig verschiedenen Aktivierungsentropien ΔS^{+} führte [45]: $\Delta G^{+} = \Delta H^{+} - T \cdot \Delta S^{+}$ ist demnach hier nur wenig temperaturabhängig.

Qualitative spektroskopische Versuche zeigen, daß das Olefin *32* sehr viel beweglicher als das Benzo-Homologe *30a* sein muß [46].

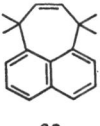

32

Aus dem Befund, daß die Dihydropleiadene bei Raumtemperatur doch noch sehr beweglich sind, folgt — ähnlich wie bei einfachen Cyclohexan-Derivaten [1,2,4a,4b] — die Frage, welche Position Substituenten in 7-Stellung des zentralen Siebenringes bevorzugen. Untersuchungen der Konformerengleichgewichte bei zahlreichen 7-substituierten 7.12-Dihydropleiadenen *33* [45,49,50)] ergaben hier ganz andere Gleichgewichtslagen als bei Cyclohexanen. So liegen z.B. 7-Phenyl-, 7-Methoxy- und 7-Acetoxy-dihydropleiaden hauptsächlich in der Konformation *A* vor, d.h. diese Reste bevorzugen eine axiale Lage. Dagegen liegt das Gleichgewicht für das 7-Methyl-Derivat überwiegend bei *B*, während eine Isopropyl-Gruppe wiederum die axiale Lage in *A* bevorzugt. Für eine eindeutige Zuordnung der Protonenresonanzsignale zu den verschiedenen Konformeren *A* und *B* wurden Modell-Betrachtungen und -Substanzen, stereospezifische Reaktionen sowie Fernkopplungen und der *Overhauser-Effekt* (NOE) zwischen axialen Protonen in 7- und 12-Stellung herangezogen [45,51)].

Zur Erklärung der im Vergleich zum Cyclohexan-System anderen Gleichgewichtslagen wird davon ausgegangen, daß dort im wesentlichen dasjenige Atom (O in OH, OCH_3 etc., C in CH_3, C_2H_5 etc.), welches direkt an den gesättigten Sechsring gebunden ist, die Unterschiede der Freien Enthalpien (*ΔG*) der Konformeren bestimmt [1,2,4)]. Bei den Dihydropleiadenen scheinen dagegen die weiteren Atome eines Substituenten von größerer Bedeutung zu sein. Für eine Isopropyl-Gruppe würde z.B. in *B* eine große sterische Hinderung zwischen den β-Methylgruppen und den benachbarten aromatischen Wasserstoffen auftreten, so daß sich hier die Lage des Gleichgewichts völlig zu *A* ($\geqslant 95\%$) verschiebt [45,49,50)].

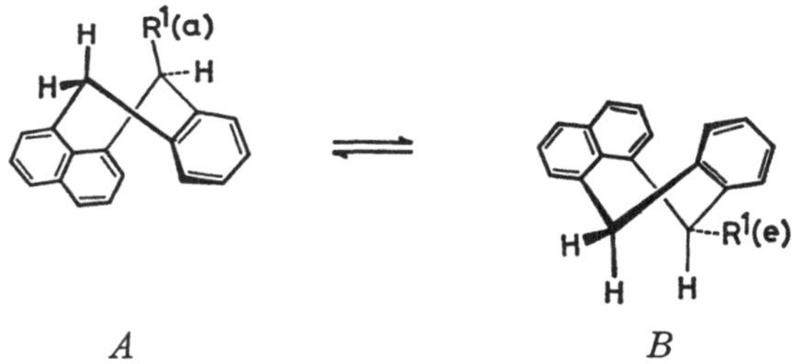

A   *B*

33

2.3. Cycloheptatriene

Das die Chemiker seit langer Zeit interessierende Cycloheptatrien/Norcaradien-Problem [52,53)] konnte ebenfalls erst in neuerer Zeit mit Hilfe physikalischer Methoden geklärt werden. IR-, Raman- [54)] und ^{1}H-NMR-spektroskopische [55)] Untersuchungen am Cycloheptatrien *(34)* selbst und an einigen Derivaten [53,56)] sprachen für das Vorliegen eines Siebenringes *(34)* und ergaben dort keine Hinweise für eine valenzautomere

Norcaradien-Struktur *(35)*. Inzwischen sind jedoch eine ganze Reihe von carbo- und auch heterocyclischen Verbindungen [52,57] (vgl. auch Abschnitt 3; S. 423, 427) bekannt geworden, denen eine bicyclische Struktur zukommt, wobei in einigen Fällen ein Gleichgewicht zwischen beiden Formen nachgewiesen werden konnte [52,57].

Beispiele sind in der carbocyclischen Reihe etwa das 7-Cyano-7-trifluor-methylcycloheptatrien *36* [58] und 7.7-Dicarbomethoxy-cycloheptatrien *37* [58a,59].

I   II

34: $R^1 = R^2 = H$ *35*: $R^1 = R^2 = H$
36: $R^1 = CN$; $R^2 = CF_3$ *37*: $R^1 = R^2 = COOCH_3$

Bei *37* konnten durch eine ausführliche ^{1}H-NMR-spektroskopische Analyse sämtliche thermodynamischen und kinetischen Parameter des Valenztautomerie-Gleichgewichtes abgeleitet werden [59]:

Lösungsmittel: CF_2Cl_2/Aceton/$CHCl_3$ (5:3:2)

ΔH ($= H_I - H_{II}$) $= 0,161 \pm 0,005$ kcal/Mol

ΔS ($= S_I - S_{II}$) $= 2,85 \pm 0,03$ e.u.

E_A (II → I) $= 7,0$ kcal/Mol, log $k_0 = 11,9$ sec^{-1}

$\Delta G^{\ddagger}_{25\,°C}$ (II → I) $= 8,2$; $\Delta G^{\ddagger}_{25\,°C}$ (I → II) $= 8,9$ kcal/Mol

Dabei stellt das Norcaradien *37 II* die enthalpieärmere Komponente des Gleichgewichtes dar, d.h. die Enthalpie des Norcaradiens liegt unter der des Triens. Auf den Entropiefaktor, der sich wegen der Starrheit des Bicyclus immer zugunsten der Trien-Form auswirkt, ist es jedoch zurückzuführen, daß bei Raumtemperatur die Differenz der Freien Enthalpien das Trien begünstigt $\Delta G_{25\,°C} = -0,688$ kcal/Mol).

In der bicyclischen Form liegen dagegen 7.7-Dicyano-norcaradien *(38)* [60], 2.5.7-Triphenyl-norcaradien *(39)* [61] sowie die Vertreter *40* [62], *41* [63] und *42* [64] vor[d),e)].

[d)] Weitere, vor allem frühere Beispiele aus der carbocyclischen Reihe finden sich in der Übersicht von G. Maier [52].

[e)] Zur Diskussion der das Gleichgewicht zwischen mono- und bicyclischer Form bestimmenden Faktoren siehe [58,59,60].

$$H_3C \quad CH_3$$

40 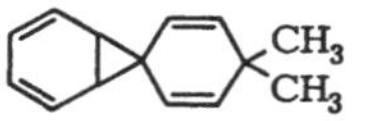   41 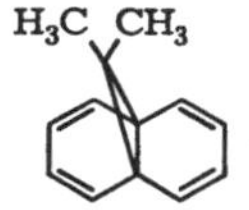   42

Für diejenigen Derivate, die als ungesättigter Siebenring *34* vorliegen, ergab sich das zusätzliche Problem der räumlichen Anordnung der sieben den Monocyclus bildenden Atome, wobei für das Cycloheptatrien selbst zunächst eine ebene und eine nicht-ebene Konformation diskutiert wurden [53,54].

Eine am p-Bromphenacylester der *7.7-Dimethyl-cycloheptatrien-3-carbonsäure 43* durchgeführte Röntgenstrukturanalyse [65] bewies für den kristallierten Zustand eindeutig das Vorliegen einer nicht-ebenen Bootkonformation: Die Atome C^7, C^3 und C^4 befinden sich oberhalb der aus C^1, C^2; C^5 und C^6 gebildeten Ebene. Die Röntgenstrukturanalyse zeigt auch eine kleine Verdrillung der C^1—C^2- und C^5—C^6-, jedoch nicht der C^3—C^4-Doppelbindung. Der große C^1—C^6-Abstand (2.42 Å) schließt hier zwar eine Norcaradien-Struktur eindeutig aus, jedoch wird eine Überlappung der π-Elektronen an C^1 und C^6 unterhalb des Bootes diskutiert; diese Orbitale sind derart gegeneinander geneigt, daß sie sich auf dieser Seite des Bootes nahe kommen können.

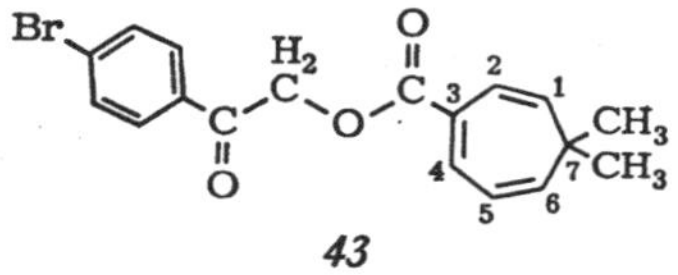

43

Für ein anderes Cycloheptatrien-Derivat — das 2-tert.-Butyl-3.7.7.-dimethylcycloheptatrien (*44*) bewiesen Conrow u. Mitarb. [66] erstmals die Wannenform in Lösung, da sie bei tiefen Temperaturen (unterhalb —86 °C) eine Nichtäquivalenz der geminalen Methylgruppen in 7-Stellung fanden. Bei Erhöhung der Aufnahmetemperatur erscheint jedoch durch Beschleunigung der Ringinversion ein Singulett doppelter Intensität für beide Methylgruppen. Für $\Delta G^{\pm}_{87\,°C}$ des Umklappvorganges ergibt sich ein Wert von etwa 9,2 kcal/Mol [67].

Damit war gezeigt, daß Cycloheptatriene bei Raumtemperatur offenbar eine schnelle Ringinversion eingehen.

Auch Cycloheptatrien *(34)* selbst zeigt — wie Anet [67] sowie Jensen und Smith [68] unabhängig voneinander fanden — unterhalb −140 °C im ^{1}H-NMR-Spektrum zwei verschiedene Absorptionen für die Methylen-protonen, die beim Erwärmen zusammenfließen. Damit ist auch für *34* in Lösung die nicht-ebene Bootkonformation eindeutig gesichert und die planare Form ausgeschlossen. Die Auswertung der Linienbreiten zeigt, daß die Energieschwelle für die Ringbewegung *34 A* $\rightleftharpoons$ *34 B* hier nur etwa 6 kcal/Mol beträgt ($\Delta G^{\ddagger}_{143\,°C} = 6{,}1$ kcal/Mol [67]; $E_a = 6{,}4 \pm 0{,}5$ kcal/Mol; $\Delta S^{\ddagger} \approx 0$ e.u. [67]; $\Delta G^{\ddagger}_{153\,°C} = 5{,}7 \pm 0{,}1$ kcal/Mol [68]).

A   *B*

34: R = H; *45*: R = D

Für das Vorliegen eines Bootes sprachen auch die transannularen (intramolekular verlaufenden) 1.5-Wasserstoff-Verschiebungen bei Cy-cloheptatrienen [69].

Interessanterweise überwiegt beim *7-Deuterocycloheptatrien 45* im Gleichgewicht das Konformere *A* mit axialem Wasserstoff und äquatori-alem Deuterium ($K = [45 A] : [45 B] = 1.41$ bei −168 °C; $\Delta H = -0{,}142 \pm$ 0,03 kcal/Mol; $\Delta S = -0{,}7 \pm 0{,}3$ e.u.), d.h. das kleinere Deuterium nimmt bevorzugt den ungünstigeren Platz (ekliptische Konformation) zwischen den C_1−H_1- und C_6−H_6-Bindungen ein. Aus den Spektren von *34* ergaben sich keine Anhaltspunkte für das Vorliegen des valenztauto-meren Norcaradiens 35 [67,68].

Eine bootförmige Struktur wurde für das Cycloheptatrien auch mit Hilfe der Elektronenbeugung (Gasphase) [70] und durch die Analyse des Mikrowellenspektrums [71] ermittelt.

Auch bei den disubstituierten Cycloheptatrienen *36 I, 37 I* sowie beim 7.7-Dimethoxycycloheptatrien *(46)* erfolgt das Umklappen des Sieben-ringes offenbar noch außerordentlich rasch, da die den geminalen Sub-stituenten dieser Triene zuzuschreibenden Signale im ^{1}H-NMR-Spektrum bzw. bei *36 I* im ^{19}F-NMR-Spektrum unterhalb −100 °C noch magneti-sche Äquivalenz zeigen [58,59,72]. Bei *36* beginnt sich unterhalb −112 °C das der Trien-Form *36 I* zugeordnete Fluorresonanzsignal langsam zu verbreitern, während unter gleichen Bedingungen die vom Norcaradien *36 II* herrührende Absorption ihre Halbwertsbreite nicht ändert [58].

Wie aus detaillierten Konformationsanalysen hervorgeht, kann die für Cycloheptatriene im Kristall, in Lösung und im Gaszustand nachgewiesene Bootstruktur jedoch hinsichtlich der Form des Bootes in Abhängigkeit von den Substituenten am Siebenring starken Veränderungen unterworfen sein.

So liefern die Elektronenbeugung [70] und das Mikrowellen-Spektrum [71] unterschiedliche Strukturwinkel α und β für den Grundkörper *34*.

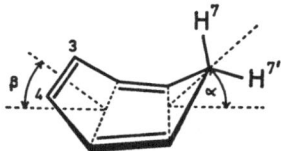

Cycloheptatrien *(34)*: $\beta = 40.5 \pm 2°$ $\alpha = 36.5 \pm 2°$ (Elektronenbeugung) [70]

                       $\beta = 29.5 \pm 4°$ $\alpha = 50 \pm 5°$ (Mikrowellen-Spektr.) [71]

   43: $\beta = 24.4°$ $\alpha = 47.9°$ (Röntgenstrukturanalyse) [65]

   49: $\beta = 0 \pm 2°$ $\alpha = 47 \pm 2°$ (Röntgenstrukturanalyse) [81]

Der Ersatz der beiden Methylenwasserstoffe in *34* durch geminale Methylgruppen in *43* führt zu einer Deformation des Bootes (Verkleinerung von β); diese Abflachung des Olefin-Teiles des Bootes dürfte auf eine sterische Wechselwirkung zwischen den π-Orbitalen der Bindung C^3—C^4 und der axial stehenden Methylgruppe am C^7 zurückzuführen sein. Für eine ^{1}H-NMR-spektroskopische Konformationsanalyse derartiger cyclischer Triene in Lösung können vicinale Kopplungskonstanten der Protonen am Siebenring, z.B. J_{23} und J_{17}, verwendet werden, da deren Größe aufgrund der von Karplus aufgefundenen Beziehungen [73] von den Dieder-Winkeln zwischen den betreffenden Bindungen abhängt (J_{23} z.B. vom Dieder-Winkel zwischen den Bindungen C^2—H^2 und C^3—H^3) [74,75]. So steigt J_{23} von etwa 5,3—5,6 Hz in *34* um mehr als 1 Hz auf 6,8—6,9 Hz in der 7.7-Dimethyl-cycloheptatrien-3-carbonsäure (*43*, H statt (p)-Br-C_6H_4-CO-CH_2-) [76], im 3.7.7-Trimethylcycloheptatrien [77] oder im 7.7-Bis-trifluormethyl-cycloheptatrien *47* [75] an, was sich nur mit der Annahme eines flacheren Bootes (Abnahme des entsprechenden Diederwinkels und damit β) interpretieren läßt [74,75].

Der Strukturwinkel β dürfte in diesen geminal disubstituierten Olefinen wie bei *43* kaum größer als 25° sein. Die Kenntnis von J_{23} erlaubt somit eine Abschätzung des Grades der Einebnung des Doppelbindungssystems. Dieses durch die Disubstitution in 7-Stellung bewirkte

Abflachen des Bootes erniedrigt offenbar die Inversionsbarriere bei *47* gegenüber *34* (6 kcal/Mol) wesentlich. Bei −185 °C konnte nämlich noch keine Nichtäquivalenz der Trifluormethylgruppen im ^{19}F-NMR-Spektrum beobachtet werden, d. h. unter diesen Bedingungen gelingt es noch nicht, die Bootform bei *47* einzufrieren [75].

Diese transannulare Abstoßung der sperrigen Trifluormethyl-Gruppen und der C^3-C^4-Doppelbindung macht sich auch im UV-Spektrum von *47* [78] bemerkbar, dessen Maximum (276—277 mμ) gegenüber anderen Cycloheptatrienen (255—269 mμ) [79] längerwellig liegt. Besonders ausgeprägt ist diese Rotverschiebung im 7.7-Bispentafluor-äthyl-cycloheptatrien *48* ($\lambda_{max} = 292$—293 mμ), in dem ein nahezu planarer Siebenring vorliegen dürfte [78].

Die Werte der vicinalen und geminalen Kopplungskonstanten J_{67}, $J_{67'}$, bzw. $J_{77'}$ von *34* lassen außerdem den Schluß zu, daß der Winkel α für den Grundkörper (in Lösung) größer als 47° [74] ist und entsprechen somit den Ergebnissen der Mikrowellen-Spektroskopie [71,74].

Zum Vergleich wurde dabei auch der kristallisierte *Cycloheptatrienmolybdän-tricarbonyl-Komplex* (*49*) [80] ^{1}H-NMR-spektroskopisch analysiert. Für *49* hatte die Röntgenstrukturanalyse des Kristalls eine interessante, vom Cycloheptatrien abweichende Struktur erbracht: Die Kohlenstoffatome des Trien-Teiles liegen in einer Ebene ($\beta = 0 \pm 2°$), über der sich auf der dem Molybdän abgewandten Seite die Methylen-Gruppe ($\alpha = 47 \pm 2°$) befindet [81].

Ausführliche NMR-spektroskopische Konformationsanalysen an mono-substituierten 7-Alkyl- und 7-Aryl-cycloheptatrienen des Typs *50* stammen von Kessler und Eu. Müller [82] sowie von Günther u. Mitarb. [83]. Aus den gefundenen chemischen Verschiebungen für H^7 sowie den Kopplungskonstanten J_{17} (zwischen H^1 und H^7) läßt sich eine starke Bevorzugung der Konformation *B* mit äquatorialem Substituenten R und axialem Wasserstoff ableiten. Für *50e* wurde ein Energieunterschied von 1,1 kcal/Mol zwischen *B* und *A* errechnet [83].

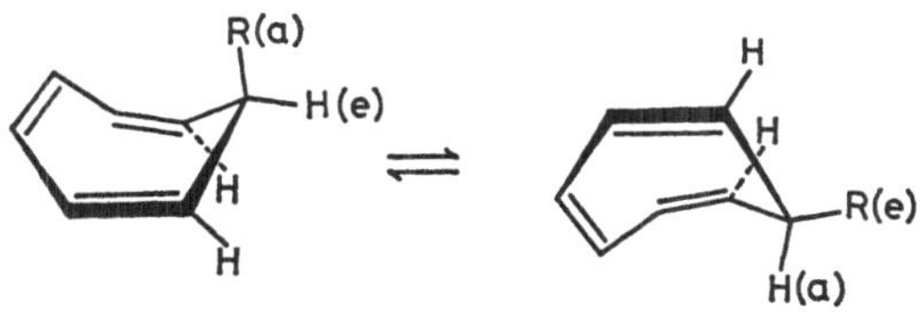

A   *50*   *B*

50a: R = CH₃	*50d*: R = C(CH₃)₃
50b: R = C₂H₅	*50e*: R = C₆H₅
50c: R = CH(CH₃)₂	*50f*: R = C(CH₃)₃; CH₃ statt H am C—1

Bei derartigen Verbindungen erscheint wie beim Cycloheptatrien selbst [67,68], bei fixierten [84] oder zahlreichen Benzo-Derivaten [88—90] und Heterocyclen [86] das axiale Methin-Proton bei höherer Feldstärke als das äquatoriale. Der Grund hierfür dürfte darin liegen, daß nach Modellbetrachtungen H^7 in B oberhalb der Ebene der C^3—C^4-Doppelbindung in einem Bereich liegt, in dem Wasserstoffatome zusätzlich abgeschirmt werden. Dieser Befund darf jedoch selbst innerhalb der Reihe der Cyclo-heptatriene nicht ohne weiteres verallgemeinert oder gar auf andere Substanzklassen übertragen werden, da sich die Verhältnisse durch Strukturabwandlungen leicht umkehren können (s. dazu Abschnitt 2.4.2, S. 415). Bei den 7.12-Dihydro-pleiadenen absorbieren äquatoriale Methin-Protonen an C^7 oder C^{12} im allgemeinen bei höherem Feld als axiale [45,49].

In der Konformation B führt die ekliptische Anordnung der Substituenten R und der C^1—H^1 bzw. C^6—H^6-Bindungen vermutlich zu einer Deformation des Bootes in der Weise, daß der Strukturwinkel α im Vergleich zum Grundkörper 34 größer wird; in A sollte dagegen aufgrund der schon auf S. 404 besprochenen Wechselwirkungen zwischen R und der C^3—C^4-Doppelbindung α umgekehrt eher kleiner werden.

Die rechnerisch ermittelten Grenzwerte für die Kopplungskonstanten J_{17} in den verschiedenen Konformationen A und B führen zu dem Ergebnis, daß α in A kleiner, in B dagegen größer als 47° ist [83].

Ein interessanter Fall stellt das von Heyd und Cupas [83a] studierte 1-Methyl-7-tert.-butyl-cycloheptatrien ($50f$) dar, bei dem im Unterschied zu $50d$ dasjenige Konformere mit axialer tert.-Butylgruppe (A) stabiler ist (Lösungsmittel CS_2). Hierfür sind wahrscheinlich sterische Behinderungen mit der Methylgruppe am C-Atom—1 in B verantwortlich. $50f$ zeigt eine für die Ringbewegung eines Monocyclus erstaunlich hohe Energieschwelle von $E_a = 18,9$ kcal/Mol (log $k_0 = 15,4$ sec^{-1}). Bei $-104\,°C$ spaltet die B zuzuordnende tert.-Butylgruppe in drei Singuletts gleicher Intensität auf (nicht-äquivalente Methylgruppen).

2.4. Benzologe Cycloheptatriene, Tropone und Tropylium-Ionen

2.4.1. Mono- und Dibenzo-cycloheptatriene (5H- und 7H-Benzo-cyclo-heptene, 5H-Dibenzo[a.d]cyclohepten) [95]

Die Annellierung von Benzolkernen an das Cycloheptatrien-Gerüst führt zu einer Erhöhung der Energiebarriere für die Inversion der Boot-Konformation, ein Phänomen, das auch für andere ungesättigte carbo- und hetero-cyclische Ringsysteme zutrifft [32,85,86]. Vermutlich ist die Versteifung darauf zurückzuführen, daß die schwerer deformierbaren o-disubstituierten Benzolringe den Valenzwinkeldeformationen, die im Übergangszustand auftreten, einen höheren Widerstand entgegensetzen als cis-substituierte Äthylene. Dieser Effekt ist zwar beim 1,4-Diphenyl- und 1.2.3.4-Tetraphenyl-7H-benzocyclohepten 51 und 52 sowie beim 1.2.3.4-Tetraphenyl-5.8-dimethyl-7H-benzocyclohepten 53 noch nicht sehr ausgeprägt ($\Delta G^{\ddagger}_{\sim -100\ °C}$ um 8 kcal/Mol) [87]; bei den zusätzlich mit Phenyl- und Methoxygruppen substituierten Verbindungen *54* und *55* macht er sich jedoch kräftig bemerkbar [88].

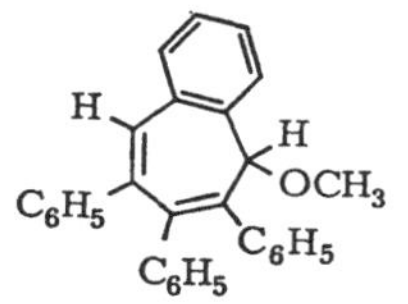

54

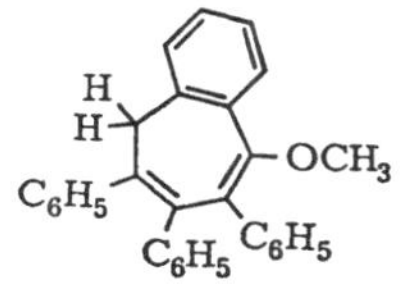

$\Delta G^{\ddagger}_{65\ °C} = 17{,}0$ kcal/Mol ($A \rightarrow B$)
$\Delta G^{\ddagger}_{65\ °C} = 16{,}8$ kcal/Mol ($B \rightarrow A$)

55

$\Delta G^{\ddagger}_{65\ °C} = 17{,}0$ kcal/Mol

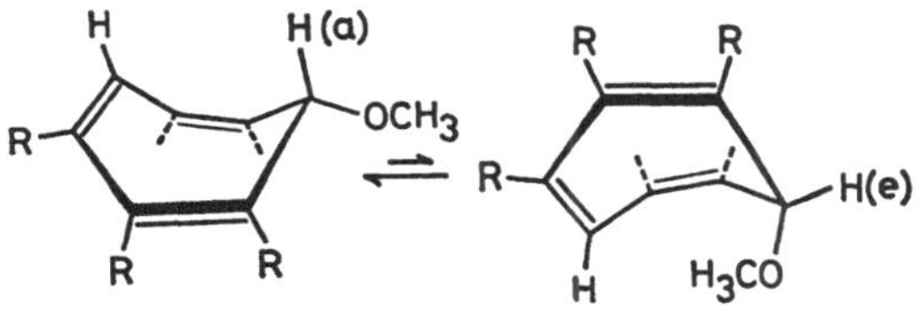

A $\qquad\qquad$ B

54

Im Tieftemperatur-Spektrum von *54* sind die Konformeren *54 A* und *54 B* nebeneinander sichtbar; die Differenz der Freien Enthalpie wurde zu $\Delta G = -0{,}25$ kcal/Mol ($-10\ °C$, $CDCl_3$) zugunsten von A mit äquatorialer Methoxylgruppe bestimmt [88].

Der Befund, daß die sehr ähnlichen Verbindungen *54* und *55* praktisch identische Energiebarrieren aufweisen, zeigt auch, daß der bei höheren Temperaturen die Äquivalenz der Methoxyle in *54* bewirkende Prozeß tatsächlich die Ringbewegung ist; ein für *54* ebenfalls zu diskutierender — unter den vorgegebenen Bedingungen ($CDCl_3$, C_4Cl_6 als Solvens) allerdings recht unwahrscheinlicher — Vorgang, nämlich die Ionisation zum entsprechenden Tropylium- und Methanolat-Ion dürfte somit eindeutig ausgeschlossen sein. Ein solcher Prozeß, welcher bei *54* ebenfalls einen Austausch der Methoxylgruppen zur Folge haben könnte, ist bei *55* nicht möglich.

54 A besitzt die bei relativ höherer Feldstärke auftretende Resonanz für das axiale Methinproton und die bei niedrigerer Feldstärke erscheinende äquatoriale Methoxylgruppe; für *54 B* liegen die Verhältnisse umgekehrt. Für konformative Zuordnungen scheint dabei die relative Lage von Methoxy- oder auch Acetoxy-Absorptionen den größeren diagnostischen Wert zu besitzen; bei den bisher untersuchten, ungesättigten und vor allem benzologen Siebenringen treten die axialen Methylprotonen dieser Substituenten bei relativ hoher Feldstärke auf [45,49–51,88, 89,90,92], da sie in dieser Konformation (z. B. in *54 B*) aufgrund ihrer Lage in einem entsprechenden Bereich oberhalb der Ebenen von $C=C$-Doppelbindungen oder Benzolringen zusätzlich abgeschirmt werden. Dagegen sind eine Reihe von Beispielen bekannt, bei denen in Abweichung vom „Normalfall" [68,82,83,86,88–90,93] axiale Methinprotonen bei niedrigerer Feldstärke als äquatoriale auftreten (vgl. auch S. 415) [45,49,91,92].

Energiebarrieren zwischen 15—18 kcal/Mol wurden auch für die Ringinversion der Dimethylacetale *56—60* ermittelt, deren ^{1}H-NMR-Spektren für die eingefrorene Bootkonformation besonders große chemische Verschiebungen (mehr als 50 Hz/60 MHz) der nicht-äquivalenten Methoxylprotonen zeigen [89].

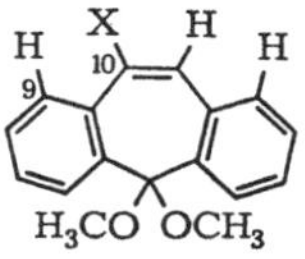

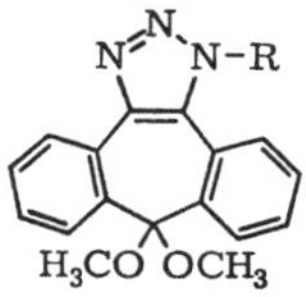

56: X=H

57: X=Br

58: R=CH$_3$

59: R=

60: R=

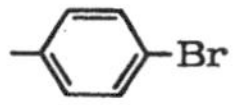

Die Erschwerung des Umklappvorganges bei *56* im Vergleich zum Cycloheptatrien *(34)* ist sowohl auf den Einfluß der geminalen Methoxylgruppen als auch auf die ankondensierten Benzolkerne zurückzuführen. Aufgrund einer soeben erschienenen Arbeit von Nogradi, Ollis und Sutherland [89a], in der die Umklappschwelle des Dibenzocycloheptatriens (5H-Dibenzo [a.d] cycloheptens [95]; *56*, H statt OCH$_3$ am C-Atom —5) zu $\Delta G^{\pm}_{85°C} = 9,2$ kcal/Mol bestimmt wurde, ist es möglich, die einzelnen Beiträge abzuschätzen: Danach steigen die Freien Enthalpien der Aktivierung von 6 kcal/Mol beim Cycloheptatrien *(34)* [67,68] durch Annellierung zweier Benzolringe auf 9,2 kcal/Mol bei der Dibenzoverbindung [89a] und schließlich auf 15 kcal/Mol beim Dimethylacetal *56* [89].

Auch bei anderen Systemen führt der Einbau von Methoxylgruppen zu einer merklichen Versteifung von nicht-ebenen Konformationen: So fanden Lansbury u. Mitarb. [45,47] beim Übergang vom 7.12-Dihydropleiaden *(30a)* zur trans-7.12-Dimethoxy-Verbindung *30e* einen Anstieg des $\Delta G^{\pm}$-Wertes von 13,6 nach 15,2 kcal/Mol. Kabuß et al. [35] beobachteten bei dem Dioxolan *61* eine um mehr als 3 kcal/Mol höhere Energieschwelle für die Ringbewegung als beim entsprechenden Kohlenwasserstoff, dem 4.4.6.6-Tetramethyl-benzocyclohepten-1.

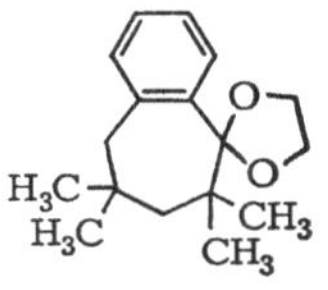

61

Wenn auch das Ansteigen der Aktivierungsenergie beim Übergang von *34* zu *56* zu einem großen Teil auf eine (durch die Substituenten und die annellierten Benzolringe bedingte) Erhöhung des Übergangszustandes (ÜZ) zurückzuführen sein dürfte, so ist doch noch ungeklärt, inwieweit auch eine Änderung des Energiegehaltes des Grundzustandes (GZ) durch zunehmende Substitution die gemessenen Umklappschwellen

$$\Delta G^{\neq} = G_{\text{ÜZ}}^{\neq} - G_{\text{GZ}}$$

beeinflußt.

Bei *56* und *57* sind noch sterische Wechselwirkungen zwischen den Substituenten X in 10-Stellung des Siebenringes [95] und dem Wasserstoff am C-Atom-9 zu berücksichtigen; der geringe Unterschied von nur etwa 1 kcal/Mol zwischen diesen beiden Verbindungen (X = H in *56* und X = Br in *57*) zeigt jedoch, daß dieser Effekt hier nicht sehr ins Gewicht fällt [89]. Bei 10-Dimethyl-hydroxymethyl-Derivaten [89a] macht er sich jedoch sehr kräftig bemerkbar.

Die Bootkonformation der *Dibenzocycloheptatriene* kann nun durch die Einführung einer semicyclischen Methylengruppe am C-Atom-5 weiter stabilisiert werden; hierdurch wird ein planarer Übergangszustand durch die zusätzliche Behinderung der Substituenten an der Doppelbindung mit den Wasserstoffen H[4] und H[6] recht ungünstig. Dies geht schon daraus hervor, daß von einigen, ^{1}H-NMR-spektroskopisch untersuchten 10-Dimethyl-hydroxymethyl-dibenzo-cycloheptatrienen *61 a* das Exomethylen-Derivat (X = C=CH$_2$) mit $\Delta G_{109\,°C}^{\neq} = 21{,}2$ kcal/Mol die höchste Umklappbarriere zeigt [89a].

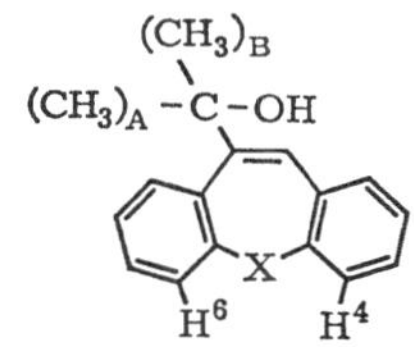

X = CH$_2$, CO, C=CH$_2$

61a

Nach Ebnöther, Jucker und Stoll [42,94] können die Konformations-Enantiomeren derartiger chiraler Methylen-dibenzocycloheptatriene (5-Methylen-5 H-dibenzo[a.d]-cyclohepten) [95] wie *62* und *63* bei Raumtemperatur getrennt werden.

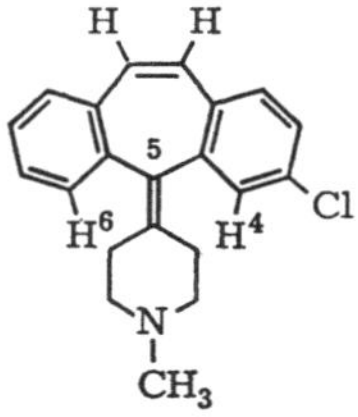

62

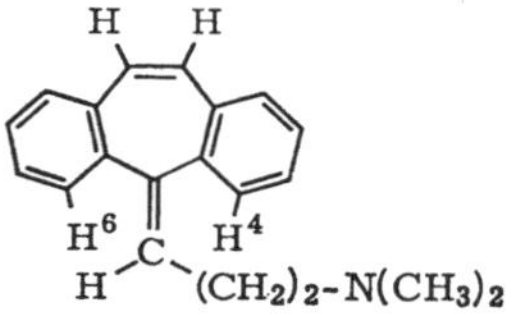

63

Die stark unterschiedlichen, polarimetrisch bestimmten Energie-schwellen für *62* ($E_A = 28$ kcal/Mol, log $k_0 = 11{,}2$ sec^{-1}) und *63* ($E_A = 21{,}9$ kcal/Mol, log $k_0 = 12{,}3$ sec^{-1}) unterstreichen die Bedeutung der Seiten-ketten an der semicyclischen Doppelbindung. So läßt sich schon *64* bei Raumtemperatur nicht mehr spalten [42].

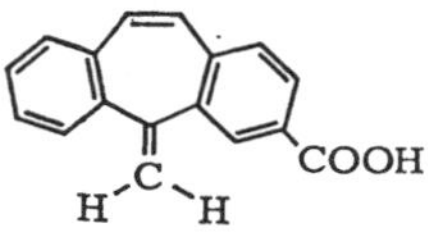

64

Andererseits sind die beiden von Schönberg, Sodtke und Praefcke [96] dargestellten syn- und anti-Isomeren des Tetrabenzo-heptafulvalens *65A* (Fp = 255 °C) und *65B* (Fp = 332 °C) [97] sehr beständig. Das bei der Synthese entstehende *65A* lagert sich erst oberhalb 200 °C in das thermodynamisch stabilere *B* um. *65B* war auch unabhängig von Berg-mann [97] erhalten worden. Die Überführung von *65A* in *65B* kann dabei

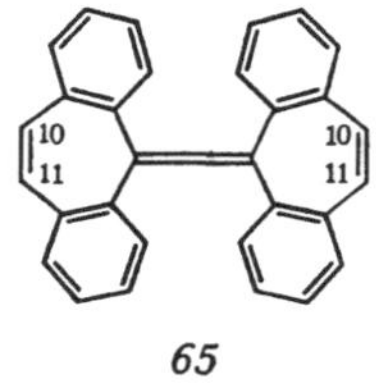

65

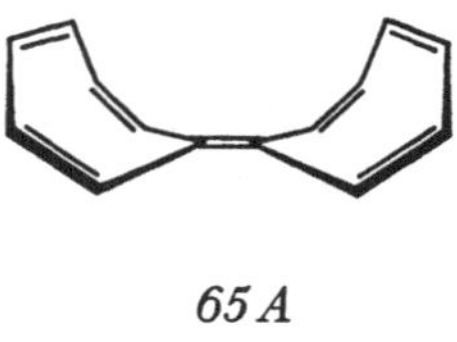

65 A

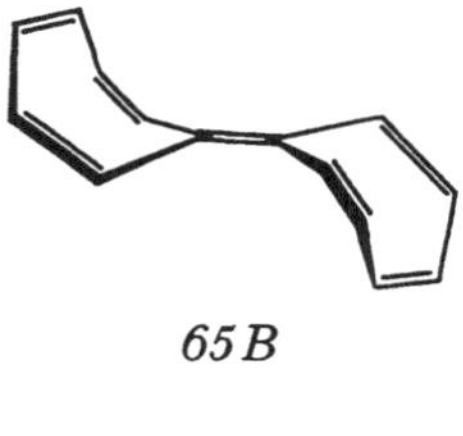

65 B

grundsätzlich auf zwei verschiedenen Wegen, nämlich Umklappen eines bootförmigen Siebenringes oder Drehung einer Molekülhälfte um die zentrale tetrasubstituierte C=C-Doppelbindung erfolgen (vgl. auch [98]).

Während somit für diese benzologen Heptafulvalene eindeutig eine nicht-ebene Bootkonformation nachgewiesen werden konnte, ergab die Röntgenstrukturanalyse des im Siebenring unsubstituierten 8.8-Dicyanoheptafulvens *66* einen planaren Bau mit allerdings alternierenden Bindungslängen [99]. 1.2-Benzoheptafulven dagegen liegt wahrscheinlich in einer schnell umklappenden Bootform vor [99a].

Ähnliche Verhältnisse bestehen bei den *Troponen* [100]: während Tropon selbst *(67a)* [101] und 2-Chlortropon *(67b)* [102] eben sind, wurde durch Röntgenstrukturanalyse des Dibenzo[b.f]tropons (5 H-Dibenzo-[a.d]cycloheptenons-5) [95] *68* und des Perchlortropons *(68a)* [103a] eine Bootform für den Kristall nachgewiesen [103]. Offenbar überwiegt beim Tropon und seinen einfachen Derivaten (z.B. Tropolone etc.) [100,104] der Gewinn an π-Stabilisierungsenergie die zu etwa 7 kcal/Mol geschätzte Ringspannung des ebenen Siebenringes [105], so daß dort die planare Konformation am energieärmsten ist [104].

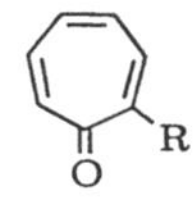

67a: R = H
67b: R = Cl (α = β = 0) [102]

Allerdings kommen Bertelli u. Mitarb. [104] neuerdings zu dem Schluß, daß der „aromatische Charakter" von Troponen und Tropolonen stark überschätzt worden sei und diese Verbindungen besser als Polyolefine beschrieben werden können. Diese Feststellung wurde aufgrund des Vergleiches von Dipolmomenten, kombiniert mit MO-Rechnungen sowie der Analyse der ^{1}H-NMR-Spektren getroffen. So ergibt sich aus den Kopplungskonstanten des Tropons, daß hier — in Einklang mit der Röntgenstrukturanalyse des 2-Chlortropons *67b* [102] — kein reguläres Heptagon, sondern ein ebener Siebenring mit alternierenden Bindungslängen vorliegt.

Entsprechendes soll auch für Tropolon-Derivate gelten [104]. (S. jedoch [164]). Der Fall eines völlig regulären Siebenringes dürfte somit am idealsten im Tropylium-Ion [100] verwirklicht sein.

2.4.2. Di- und Tribenzocycloheptatriene (5 H-Dibenzo[a.c]cycloheptene und 9 H-Tribenzo[a.c.e]cycloheptene) [95)]

Während die Einführung eines Triazol-Ringes bei *58—60* in die 10- und 11-Stellung des Dibenzocycloheptatrien-(5 H-Dibenzo[a.d]cyclohepten-)-Systems einen nur mäßig erschwerenden Einfluß auf das Umklappen hat, steigt die Energieschwelle für die Ringbewegung im Tribenzocycloheptatrienon-dimethylacetal *69 a* auf mehr als 23 kcal/Mol an [89)] und wurde somit gegenüber dem nur um einen ankondensierten Benzo-Ring ärmeren *56* um mehr als 8 kcal/Mol erhöht. *69 a* zeigt im ^{1}H-NMR-Spektrum stets zwei verschiedene, auch bei 180° noch nicht merklich verbreiterte Singulett-Resonanzen für die Methoxylgruppen. In Einklang mit den Vorstellungen von Stiles [106)] und Heilbronner [107)] darf diese bemerkenswerte Versteifung von *69 a* auf die gegenseitige räumliche Behinderung von vier Wasserstoffen in ortho-Stellung zum Siebenring zurückgeführt werden. Der eingeebnete Übergangszustand wird dadurch im Verhältnis zum Grundzustand stärker angehoben als bei *56*.

69 a: R = H; X = C(OCH$_3$)$_2$	$\Delta G^{\ddagger}_{>180\ °C} > 23$ kcal/Mol
69 b: R = H; X = CH$_2$	$\Delta G^{\ddagger}_{202\ °C} = 24{,}0$ kcal/Mol
69 c: R = CH$_2$CH$_3$; X = CH$_2$	$\Delta G^{\ddagger}_{>200\ °C} > 27{,}7$ kcal/Mol
69 d: R = CH$_2$CH$_3$; X = CO	$\Delta G^{\ddagger}_{125\ °C} = 20{,}0$ kcal/Mol

Auch die neuerdings spektroskopisch studierten Tribenzocycloheptatriene *69 b—69 c* zeigen ähnlich hohe Energieschwellen [89 a)].

Obwohl in dieser Reihe das 1-Äthyl-tribenzotropon (*69 d*) mit $\Delta G^{\ddagger}_{125\ °C} = 20{,}0$ kcal/Mol am beweglichsten ist, scheint dem planaren Übergangszustand der Ring-inversion keine besonders große Stabilisierung durch einen zusätzlichen Gewinn an Mesomerieenergie bei der Einebnung zuzukommen. Unter gewissen Annahmen wurde nämlich für *69 d* bei Berücksichtigung aller sterischen Wechselwirkungen (*nonbonded interaction energies*) eine Inversionsbarriere von ca. 30 kcal/Mol berechnet. Danach läge der Gewinn an zusätzlicher Mesomerieenergie in einem planaren Tribenzotropon nur in der Größenordnung von 10 kcal/Mol [89 a)].

Die von Sutherland und Ramsay [39)] untersuchten *1.2.3.4-Dibenzocycloheptatriene* (5 H-Dibenzo[a.c]cycloheptene) [95)], bei denen eine ähnliche Wechselwirkung von nur zwei orthoständigen Wasserstoffen vorliegt, zeigen $\Delta G^{\ddagger}$-Werte, die mit 18—19 kcal/Mol zwischen denen von 56 und 69 liegen.

Die obige Deutung wird außerdem auch dadurch gestützt, daß das Pyrazin-Derivat 1 eine erheblich niedrigere Umklappschwelle ($\Delta G^{\ddagger}_{120\,°C}$ = 19,8 ± 0,3 kcal/Mol, E_A = 21,4 ± 1,3 kcal/Mol; log k_0 = 13,9 ± 0,7 sec^{-1}) als 69a zeigt. Demnach erfolgt hier das für die Ringinversion notwendige Vorbeigleiten der Stickstoffatome mit freiem Elektronenpaar an den C—H-Gruppen der benachbarten Benzolringe merklich leichter als der entsprechende Vorgang mit vier C—H-Gruppen in 69a [89].

Die diskutierten, NMR-spektroskopisch erhaltenen Ergebnisse über die bemerkenswerte Versteifung der Bootkonformation bei Tribenzocycloheptatrienen führten 1964 zu den folgenden *Voraussagen* [89]:

Chirale Derivate des Tribenzocycloheptatriens sollten bei Raumtemperatur in stabile Konformationsenantiomere spaltbar sein;

für Abkömmlinge, die am tetraedrischen Kohlenstoffatom in der 9-Stellung zwei verschiedene Substituenten tragen, sollten isolierbare Konformationsisomere zu erwarten sein.

Die erste Vorhersage konnte bei der 9 H-Tribenzo[a.c.e]cycloheptenyliden-9-essigsäure 70 [108] verwirklicht werden (Racematspaltung über diastereomere Brucin-Salze), deren recht beständige Enantiomere A und B erst zwischen 100 °C und 140 °C meßbar langsam racemisieren.

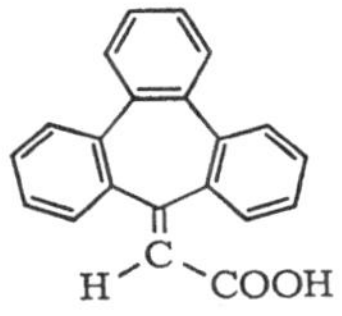

70

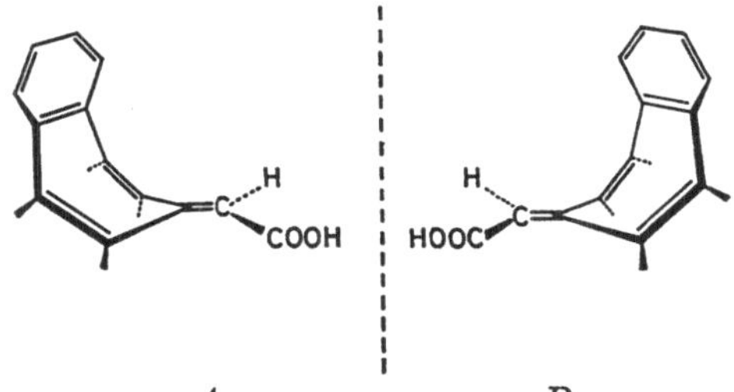

A   B

70

Aufgrund der gefundenen Aktivierungsparameter ($\Delta G^{\ddagger}_{110 °C} = 32$ kcal/Mol; $E_A = 30$ (± 2) kcal/Mol; $\Delta S^{\ddagger}_{110 °C} = -7$ (± 5) e.u.; log $k_0 = 11{,}7$ sec^{-1}) ist es wahrscheinlich, daß die Racemisierung durch das Umklappen des Siebenringes erfolgt. Allerdings kann nicht mit letzter Sicherheit ausgeschlossen werden, daß ein anderer Vorgang — nämlich die Rotation um die semicyclische Doppelbindung — an der Racemisierung beteiligt ist [108].

Die Energiebarriere liegt bei dem Tribenzo-Derivat um $\Delta\Delta G^{\ddagger}_{110°} = (\Delta G^{\ddagger}_{110 °C}$ *(70)* $- \Delta G^{\ddagger}_{110 °C}$ *(63)* $= 32 - 23) = 9$ kcal/Mol höher als bei der Dibenzo-Verbindung *63*. Geht man von der Annahme aus, daß sich durch die Carboxyl- und Dimethylaminoäthyl-Gruppe in *70* bzw. *63* bewirkten sterischen Hinderungen nicht stark unterscheiden [109], so ist diese Differenz in erster Linie wieder auf die durch den dritten Benzolring in *70* bewirkte Versteifung zurückzuführen.

Auch stabile konformationsisomere Tribenzocycloheptatriene sind dargestellt worden. So liefert 9-Hydroxy-9 H-tribenzo[a.c.e]cyclohepten *(71)* mit vermutlich äquatorialer Hydroxylgruppe *(A)* [90,110] bei der Verätherung mit Methanol/Schwefelsäure einen Methyläther *72*, in dem laut ^{1}H-NMR-Spektrum nur das Konformere mit äquatorialer Methoxylgruppe vorliegt. Nach dem Erhitzen in Hexachlorbutadien findet man im Spektrum zu etwa 10% auch das zweite Konformere mit axialem Methoxyl *(B)* [90,111].

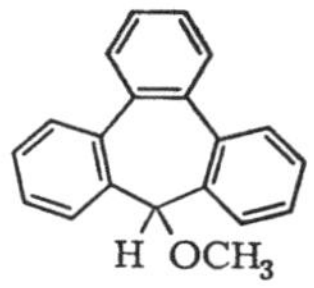

72

In der konformativ besonders beständigen Tetraphenyltribenzocycloheptatrien-(1.2.3.4-Tetraphenyl-9 H-tribenzo[a.c.e]cyclohepten-) [95]-Reihe [90,91] ist es neuerdings gelungen, in mehreren Fällen *(73, 77* und *78)* die Konformationsisomeren *A* und *B* rein zu erhalten; bei *74, 75* und *76* wurden die äquatorialen Vertreter rein dargestellt [112].

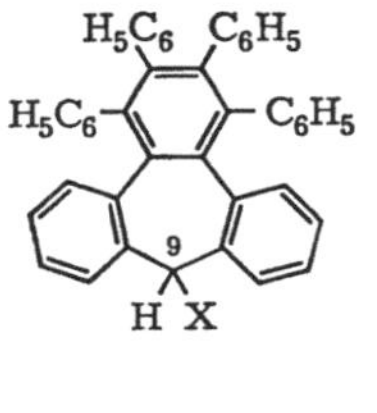

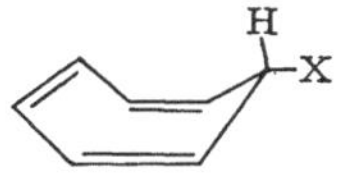

A	B

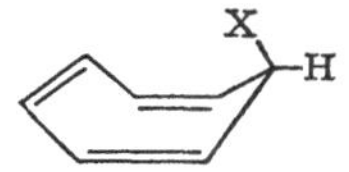

73A: X = OH	*73B*: X = OH
74A: X = OSCl	
	O
75A: X = Cl	*75B*: X = Cl
76A: X = OTs	
77A: X = OCH$_3$	*77B*: X = OCH$_3$
78A: X = OAc	*78B*: X = OAc
84: X = H	*85B*: X = Br

(vgl. auch Abschnitt 2.4.3)

Die Isomeren des Typs A und B sind in siedendem Benzol noch stabil, können jedoch in Xylol oder Diäthylenglykoldimethyläther bei 140° äquilibriert werden; bei *73*, *77* und *78* sind die Konformeren A mit äquatorialem Substituenten bevorzugt [90]. *73B* zeigt im ^{1}H-NMR-Spektrum in DMSO-D$_6$ (langsamer Hydroxylprotonen-Austausch) das verglichen mit *73A* bei höherer Feldstärke erscheinende Hydroxylproton; im IR-Spektrum von *73B* treten dagegen die OH-Valenzschwingungen aufgrund von intramolekularen OH-π-Elektronen-Wechselwirkungen bei niederen Wellenzahlen auf. Beide Effekte dürften darauf zurückzuführen sein, daß in *73B* die axiale Hydroxylgruppe oberhalb der Benzolringe des Tribenzocycloheptatrien-Systems zu liegen kommt.

Aus dem gleichen Grund erscheinen bei *77B* und *78B* die Methylsignale der axialen Methoxy- und Acetoxy-Gruppen bei deutlich höherer Feldstärke als bei *77A* und *78A*. Dagegen treten bei allen bisher untersuchten Konformeren A im Gegensatz zu *72* und in Abweichung vom „Normalfall" (siehe Seite 408) die axialen Methinprotonen am C-Atom-9 im ^{1}H-NMR-Spektrum bei tieferer Feldstärke auf als die äquatorialen in B. Ähnliche relative chemische Verschiebungen wie bei *77A* und *77B* finden sich auch in den von Childs und Winstein [92] untersuchten epimeren Dibenzohomocycloheptatrienylmethyläthern *79A* und *79B*.

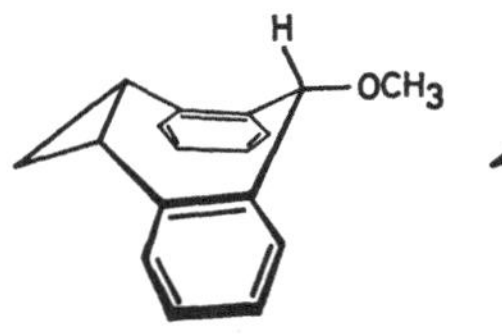

79A 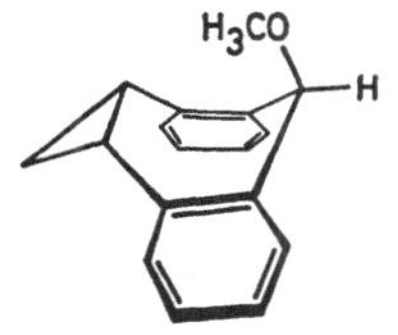   *79B*

Diese Befunde zeigen einmal mehr, daß offenbar der gegenüberliegende Benzolring in *72* bzw. die gegenüberliegende Doppelbindung in *54* die Lage der Resonanz axialer Methinprotonen wesentlich bestimmt: so erscheinen H_a in *72* und in *54* bei höherem Feld als H_e; bei *79*, in dem sich an dieser Stelle ein Cyclopropyl-Ring befindet, kehren sich die Verhältnisse (wie auch bei den Dihydropleiadenen) um. Da das gleiche für das Tetraphenyltribenzocycloheptatrien-System gilt ($H_a < H_e$ bei *73* und *77*), muß man in Einklang mit ähnlichen Befunden von Cristol und Nachtigall [112] annehmen, daß der phenyl-substituierte Benzolring in diesen Verbindungen eine geringere Abschirmung als der unsubstituierte in *72* besitzt. Daher erscheint auch die axiale Methoxylresonanz in *77B* ($\tau_{OCH_3} = 6{,}52$) bei niedrigerer Feldstärke als bei *72B* ($\tau_{OCH_3} = 7{,}22$). Somit kann auch hier gesagt werden, daß bei der Zuordnung von Methinresonanzen zu bestimmten Konformeren Vorsicht geboten ist; tunlichst werden hierfür andere Informationen [45,51,92] herangezogen.

2.4.3. Konformativ bedingte Reaktivitätsunterschiede bei Tribenzocycloheptatrienen (9 H-Tribenzo[a.c.e]cycloheptenen)

Aus der relativ hohen Energieschwelle für die Ringinversion und damit auch für die Einebnung folgt die Frage, inwieweit sich diese Versteifung, d. h. Stabilisierung der Bootkonformation, auf die chemische Reaktivität dieser Verbindungen auswirkt [90,91]. Die mögliche Isolierung von Konformationsisomeren [91] in der Tetraphenyl-Reihe gestattet es außerdem, nach konformativ bedingten Reaktivitätsunterschieden zwischen *A* und *B* zu suchen.

Aus diesem Grund wurde die Bildung substituierter Tribenzo-tropylium-Ionen eingehend untersucht. *Heilbronner* u. Mitarb. [107,113] hatten bereits 1960 die beim Tribenzo-tropylium-Ion *81 a* auftretende Abweichung von der — für andere benzologe Tropylium-Ionen gut erfüllten — Korrelation zwischen dem gemessenen $pK_R\oplus$-Wert und der Atomlokalisierungsenergie auf eine sterische Einschränkung der Koplanarität in *81 a* zurückgeführt. Später konnte am Beispiel des 2-Carboxy-tribenzo-tropylium-Ions *81 b* ($\lambda_{max} = 412$, 525 mμ in H_2SO_4) gezeigt werden, daß die Bildung derartiger Kationen aus den entsprechenden Pseudobasen bei Raumtemperatur in konz. Schwefelsäure langsam erfolgt ($t_{1/2} \approx 25$ min) [90,111].

Zu dem schwierigen, noch nicht endgültig gelösten Problem der Konformation von *Tribenzotropylium-Ionen* liegen folgende Ergebnisse vor: Die optisch-aktive Carbonsäure *80b* (äquatoriale Hydroxylgruppe) [110] liefert mit Methanol/Schwefelsäure bei Raumtemperatur einen Methoxy-tribenzocycloheptatrien-carbonsäuremethylester *82* (*80b*; OCH_3 statt OH; $R^2 = COOCH_3$; Konformeren-Verhältnis A:B = > 90: < 10), der noch optisch aktiv ist. Die Bildung von *82* verläuft unter teilweiser Retention, begleitet von beträchtlicher Racemisierung [111].

Demnach kann *82* nicht vollständig über ein symmetrisch solvatisiertes, koplanares — und damit zwangsläufig achirales 2-Carboxy-tribenzotropylium-Ion *81b* gebildet werden. Die Resultate lassen sich dagegen besser mit der Annahme eines nicht völlig eingeebneten, bootförmigen und damit noch chiralen Kations *81b* erklären. Allerdings muß die Frage, ob für die Bildung des optisch aktiven Esters *82* nicht nur asymmetrische Solvatationseffekte des Carbonium-Ions *81b* verantwortlich sind, noch geklärt werden [111,114].

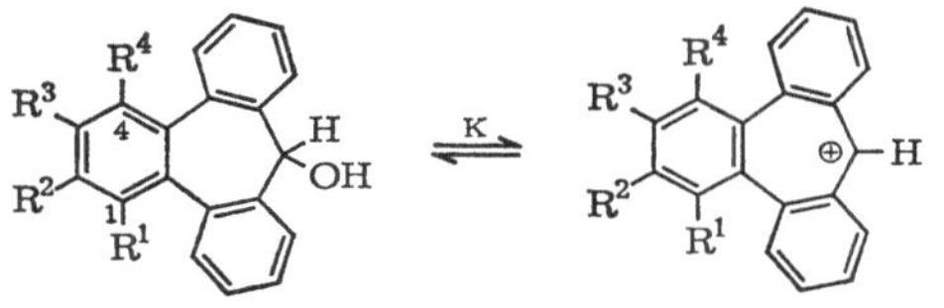

a: $R^1 = R^2 = R^3 = R^4 = H$
b: $R^1 = H$; $R^2 = COOH$; $R^3 = R^4 = H$
c: $R^1 = CH_3$; $R^2 = R^3 = R^4 = H$
d: $R^1 = R^4 = CH_3$; $R^2 = R^3 = H$

Führt man in die für die intramolekulare Beweglichkeit des bootförmigen Tribenzocycloheptatrien-Systems entscheidenden 1- und 4-Positionen größere Substituenten (z. B. Methyl- oder Phenyl-Gruppen) ein, so gelangt man zu den Carbinolen *73, 80c* und *80d*. Durch Hydridreduktion der entsprechenden Ketone entstehen praktisch ausschließlich die Konformeren *A* mit äquatorialer Hydroxylgruppe [90,91,92,110].

Diese Pseudobasen erweisen sich bei Reaktionen, die zu dem Carbonium-Ion führen sollten, als außerordentlich reaktionsträge: So zeigen *73A* sowie *80c* und *80d* (in der Konformation *A*) — im Gegensatz zu *80a* und *80b* — mit konz. Schwefelsäure keine Halochromie und lassen sich auch mit Methanol/Schwefelsäure nicht mehr veräthern [90,91]. Somit wird durch die (sich vier Kohlenstoffatome vom Reaktionszentrum entfernt befindenden) Methyl- oder Phenylgruppen die Reaktivität dieser „*Cycloheptatrienole*" im Vergleich zu *80a* und *80b* stark beeinflußt, da sich die Substituenten in den für die konformative Beweglichkeit des Siebenringes entscheidenden Positionen befinden.

Bemerkenswerterweise bleibt auch die Umsetzung von *73A* mit Thionylchlorid auf der Stufe des Chlorsulfits *74A* stehen [91]; *74A* ist thermisch recht stabil und zersetzt sich in Lösung erst oberhalb von

100 °C zu einem Gemisch der konformationsisomeren Chloride *75A* und *75B* [f].

Im Gegensatz zu der gegenüber Solvolysen inerten äquatorialen Chlorverbindung *75A* geht das axiale Chlorid *75B* mit Nucleophilen leicht Substitutionsreaktionen ein, die unter ganz bevorzugter Erhaltung der Konformation ablaufen (*75B → 73B*, *75B → 77B*, *75B → 78B*). Im Unterschied zu *73A* geht das axiale Carbinol *73B* mit Methanol/Schwefelsäure ebenfalls unter Konformationserhaltung in *77B* über. Insgesamt erweisen sich somit Verbindungen des Typs *73A—75A* (äquatoriales X) bei Umsetzungen, die über das entsprechende „*Tetraphenyltribenzotropylium-Ion*" *81* verlaufen sollten, als recht reaktionsträge, während die Derivate des Typs *B* ohne Schwierigkeiten unter Retention reagieren. Offensichtlich sind bei dem starren Tetraphenyl-tribenzocycloheptatrien-System die Voraussetzungen für die Stabilisierung eines Carbonium-Ions (oder zumindest des zu ihm führenden Übergangszustandes) durch die π-Elektronen benachbarter Benzolringe bei Ionisation einer axialen C—X-Bindung in *B* wesentlich günstiger als bei Ionisation einer äquatorialen in *A*. Wie Modellbetrachtungen zeigen (siehe Abb. 3), kann ein in die ursprünglich axiale Bindungsrichtung weisendes Orbital — bei Annahme einer gewissen Abflachung des Siebenringes, d. h. vor allem Verkleinerung des Strukturwinkels α — mit den π-Elektronen der in Frage kommenden Benzolringe in Wechselwirkung treten.

Umgekehrt sind die Voraussetzungen für eine Orbitalüberlappung bei Ionisation einer äquatorialen C—X-Bindung sehr viel ungünstiger, da diese Bindungsrichtung und die Richtung der p_z-Achse der π-Elektronen der Benzolringe eher senkrecht als parallel zueinander stehen.

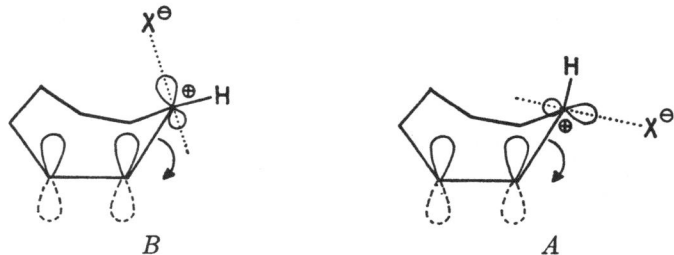

Abb. 3. Schematische Darstellung der C—X-Ionisation in den Konformeren B und A.

Hierbei stellt sich auch noch die interessante Frage, ob nicht nach Ionisation der axialen C—X-Bindung das leere Orbital am C-Atom-9 des

---

[f] Unter den Thermolysebedingungen von *74A* ist eine thermische Äquilibrierung der Konformeren *A* und *B* von *74* und *75* möglich.

„Tetraphenyl-tribenzotropylium-Ions" eine gewisse s-Beimischung behält [115]. Dazu müßte allerdings geklärt werden, ob der für die Abweichung von der trigonal-ebenen Geometrie (sp^2-Hybridisierung der drei restlichen Bindungen) erforderliche Energieaufwand [116] durch Gewinn an π-Stabilisierungsenergie kompensiert oder überkompensiert werden kann.

Andere stereochemische Verhältnisse liegen bei dem flexibleren *Cyclopropa-dibenzocycloheptadien-* (*10.11-Methylen-10.11-dihydro-5 H-dibenzo[a.d]cyclohepten*)[95])-*System 79* vor: Dort kann sich das durch Delokalisation und Homokonjugation stabilisierte Dibenzohomotropylium-Ion *83*, dessen ^{1}H-NMR-Spektrum in H_2SO_4 aufgenommen werden konnte, aus den beiden epimeren Acetaten *79A* und *79B* (OAc statt OCH$_3$) bilden [92]. Dabei hydrolysiert das äquatoriale Acetat *79A* (OAc statt OCH$_3$) 10^2mal schneller als das axiale *79B* (OAc statt OCH$_3$), wobei sich aus beiden Verbindungen bevorzugt das äquatoriale Carbinol und weniger das axiale bildet [92].

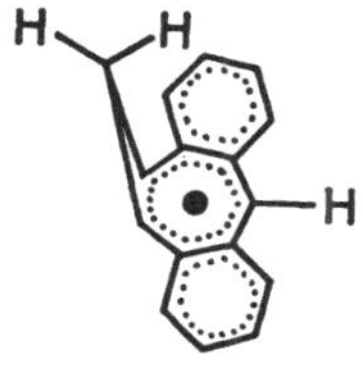

83

Auch bei anderen Reaktionstypen wurde eine bevorzugte Ablösung axialer Substituenten in konformativ fixierten Systemen gefunden: So liefert die Peroxid-katalysierte Bromierung des Tetraphenyl-tribenzocycloheptatriens *84* mit N-Brom-succinimid in CCl$_4$ zu mehr als 80% das axiale Bromid *85B* (X = Br) [117]. Lansbury u. Mitarb. [45] fanden, daß im 7-Isopropyliden-7.12-dihydropleiaden *31b* ganz bevorzugt das axiale Proton am C-Atom-12 einem H/D-Austausch mit Kalium-tert.-butanolat in DMSO-D$_6$ unterliegt.

Somit sind die beiden diastereotopen Wasserstoffe an der Methylengruppe der Kohlenwasserstoffe *31b* und *84* aufgrund unterschiedlicher konformativ bedingter π-Stabilisierungsmöglichkeiten für Zwischenstufen (oder zumindest der zu ihnen führenden Übergangszustände) mit dreibindigem Kohlenstoff in ihrem reaktiven Verhalten sehr verschieden: Dabei ähneln die axialen Wasserstoffe mehr dem Methin-Wasserstoff des Triphenylmethans, die äquatorialen dagegen zeigen eine Reaktivität, die eher mit derjenigen des Brückenkopf-Wasserstoffatoms im Triptycen vergleichbar ist [118].

3. Heterocyclische Verbindungen

Zahlreiche heterocyclische Analoga der im Abschnitt 2 besprochenen Carbocyclen wurden untersucht. Zwar ergaben sich dabei meist ähnliche konformative Verhältnisse wie bei den nur Kohlenstoff enthaltenden Ringsystemen, jedoch bedingt die Einführung von Heteroatomen aufgrund unterschiedlicher sterischer Parameter (C-X-Bindungsabstände, C—X—C-Bindungswinkel, Torsionsschwellen und -winkel bei C—X-Bindungen etc.) oft eine deutliche Veränderung der Konformation sowie der Energiebarriere für die Ringbewegung im Vergleich zum entsprechenden Carbocyclus.

3.1. Stickstoff-Heterocyclen

3.1.1. Gesättigte Ringsysteme

Beim N-Methyl-homopiperidin (N-Methyl-hexahydro-azepin) *86* ist zu erwarten, daß die Pseudorotation und damit die Ringinversion ebenso wie beim Cycloheptan eine sehr kleine Aktivierungsenergie besitzt.

Die im ^{1}H-NMR-Spektrum von *86* gefundene Temperaturabhängigkeit der zum Stickstoff α-ständigen Methylenprotonen ist auf die Inversion des Stickstoffs zurückzuführen, die bei sehr tiefen Temperaturen ($T_c = -100$ °C) „eingefroren" werden kann. Für diesen Prozeß ergibt sich eine Freie Enthalpie der Aktivierung $\Delta G^{\pm}$ von etwa 7 kcal/Mol [119]. Über die Ringinversionsschwelle wurden keine Angaben gemacht.

3.1.2. Azepine, Diazepine, ihre Dihydro- und Monobenzo-Derivate

Die Konformationen und Ringbewegungen der Azepin-Abkömmlinge *87* [120], *88—89* [121], *90* [122], *91—93* [121], *94—95* [123] und *96—97* [124] wurden ^{1}H-NMR-spektroskopisch untersucht.

Dabei wurde die Temperaturabhängigkeit der Signale der diastereotopen Methylenprotonen im Ring oder im Fall von *92*, *94c—94e* der ebenfalls diastereotopen Methylen-H oder Methylgruppen von Äthoxyl-, Äthyl-, Isopropyl- oder Benzyl-Substituenten am Ring (in den Formeln mit A und B bezeichnet) ausgewertet. Diese Möglichkeit der Beobachtung Ringsubstituenten ist vor allem bei *92* von Bedeutung, da dort der zentrale ungesättigte Siebenring über kein sp^3-hybridisiertes Ringkohlenstoffatom mit geeigneten Resten verfügt, deren chemische Umgebung sich bei der Ringbewegung verändert. Nun zeigt aber *92* bei —20 °C ein ABX_3-Spektrum für die Äthoxylgruppe. Hier wirkt also offensichtlich der Siebenring als Chiralitätszentrum, durch das die beiden Methylenprotonen magnetisch nicht-äquivalent werden. Bei Temperaturerhöhung verschwindet die Asymmetrie im Zeitmittel durch einen schnellen Vorgang, sehr wahrscheinlich durch die Ringinversion des Siebenringes [121].

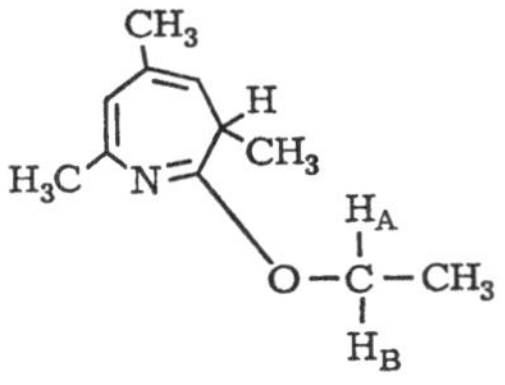

87

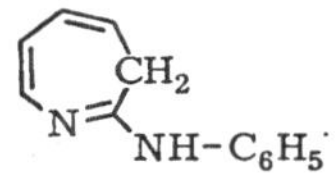

88

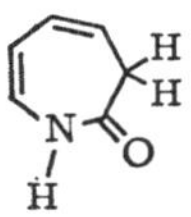

89

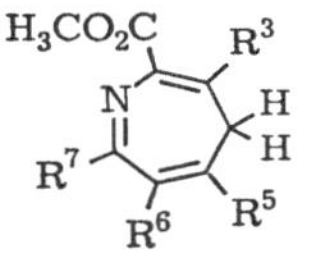

90

a: $R^3=R^5=H$; $R^6=R^7=COOCH_3$
b: $R^3=CH_3$; $R^5=H$; $R^6=R^7=C_6H_5$
c: $R^3=H$; $R^5=R^6=R^7=CH_3$

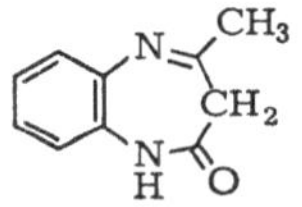

91

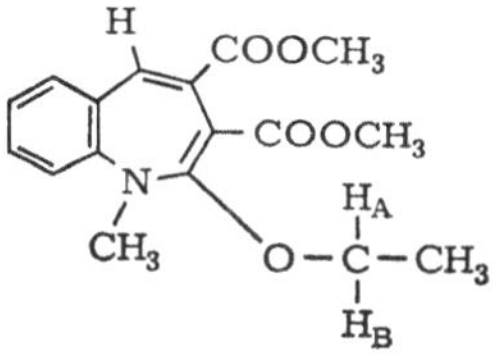

92

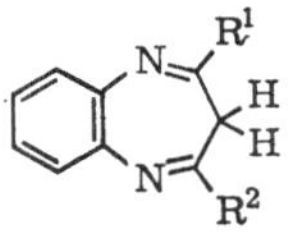

93

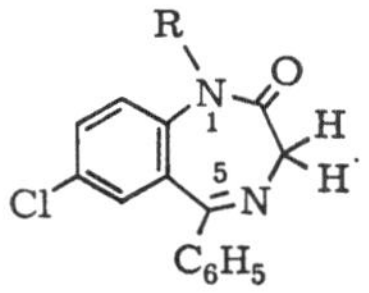

94

a: $R^1=R^2=C_6H_5$
b: $R^1=CH_3$; $R^2=C_6H_5$
c: $R^1=R^2=CH_3$
d: $R^1=R^2=-CH=CH-C_6H_5$

a: $R=H$
b: $R=CH_3$
c: $R=CH_AH_BCH_3$
d: $R=CH(CH_3)_A(CH_3)_B$
e: $R=CH_AH_B-C_6H_5$

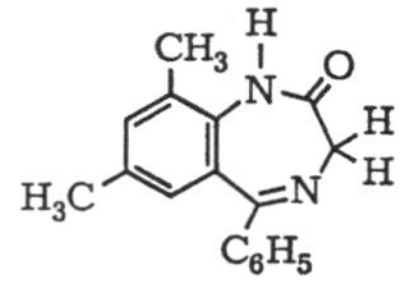

95

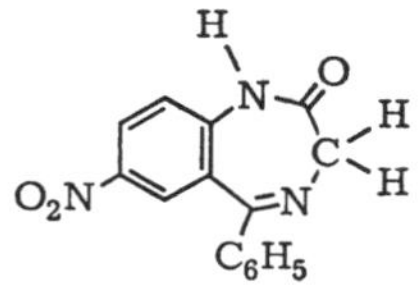

96

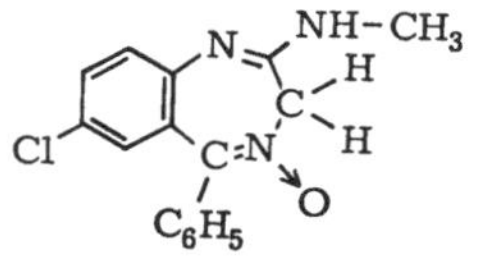

97

Die freien Enthalpien der Aktivierung $\Delta G^{\pm}$ für *88* ($T_c = -55\,°C$) und *89* ($T_c \sim -90\,°C$) liegen bei 10,2 bzw. ca. 8,5 kcal/Mol [121]), während sie bei den Monobenzo-Derivaten auf ca. 12—19 kcal/Mol ansteigen [121, 123,124]. Die in mehreren Fällen bestimmten Aktivierungsentropien $\Delta S^{\pm}$ waren negativ (bis zu (-10)—(-15)) e.u. [123,124]. Diese Daten wurden durchweg der Umklappbewegung des nichtebenen Siebenringes zugeschrieben; eine für *92* ebenfalls diskutierte Stickstoffinversion konnte aufgrund der relativ hohen Energieschwelle ($\Delta G^{\pm}_{35\,°C} = 15,3$ kcal/Mol) mit ziemlicher Sicherheit ausgeschlossen werden. Verbindungen mit drei Doppelbindungen im Siebenring (*87, 88, 90, 92, 93* und *97*) dürften dabei wie Cycloheptatriene *34* (s.S. 404) in einer Bootkonformation vorliegen, während für *89, 91, 94—96* wegen der möglichen Amid-Mesomerie eine Mittelstellung zwischen den Cycloheptadienen mit ihrer verdrillten (oder Pseudo-) [123] Bootkonformation (vgl. *18* und *20*) und den Cycloheptatrienen anzunehmen ist. Im Einklang mit den ¹H-NMR-spektroskopischen Resultaten bei monocyclischen Azepinen ergibt auch die Röntgenstrukturanalyse des N-(p-Brombenzolsulfonyl)-1 H-azepins (*98*) das Vorliegen einer Bootkonformation ($\alpha = 23°$, $\beta = 25°$) mit lokalisierten Doppelbindungen im Siebenring [125].

Der bemerkenswerte Befund, daß *2-Anilino-3 H-azepin* (Dibenzamil) *88* eine um 4 kcal/Mol höhere Umklappschwelle als Cycloheptatrien aufweist, ist noch nicht endgültig geklärt. Von einem Azacycloheptatrien könnte man nämlich eine größere Beweglichkeit als von der Stammverbindung erwarten [121]. Möglicherweise ist der versteifende Effekt auf die vorhandene Anilino-Gruppe (starke Senkung des Grundzustandes durch Amidin-Mesomerie?) zurückzuführen [121].

Von Linscheid und Lehn [123] wurden die verschiedenen Beiträge zur Höhe der Energiebarrieren bei den Dihydro-benzoazepinonen *94* folgendermaßen abgeschätzt:

9 kcal/Mol für das „Grundgerüst" *91* [121],

3 kcal/Mol durch sterische Wechselwirkungen des Benzo-Ringes mit der zusätzlichen Phenylgruppe am C-Atom 5,

4—6 kcal/Mol durch analoge Behinderungen der Reste R am N-Atom 1.

Während in den Spektren der bislang besprochenen Verbindungen *87—98* keine Anzeichen für *Valenztautomere mit dreigliedrigem Ring* gefunden wurden, liegen nach eingehenden Untersuchungen von Maier u. Mitarb. [52,126,127], Sauer u. Mitarb. [158,159,160] sowie Battiste und Barton [128] zahlreiche Derivate des 3.4-Diazacycloheptatriens (5 H-1.2-Diazepins), z.B. *99*, ganz bevorzugt in der bicyclischen 3.4-Diazanorcaradien-Form *100* vor. Die Begünstigung von *100* ist dabei auf die beiden Stickstoffatome zurückzuführen, die im Bicyclus an zwei C=N-Bindungen beteiligt sind, während in *99* die energiereiche N=N-Gruppe vorliegt. Die Konzentration an monocyclischem Trien liegt dabei oft unter der Nachweisgrenze [127]. Ein indirekter Nachweis für *99* gelang jedoch Maier und Heep [52,127] durch ^{1}H-NMR-spektroskopische Untersuchung des valenztautomeren Gleichgewichts zwischen *cis*- und *trans-100* [E_a (*cis-100* → *trans-100* und *trans-100* → *cis-100*) ~ 22,5 kcal/Mol; $\Delta S^{\neq}$ ~ —4 e. u.], das sehr wahrscheinlich über das Trien *99* verläuft [127].

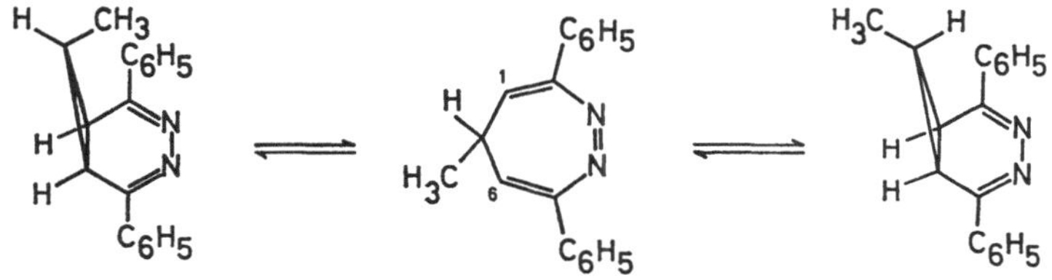

cis—*100*   *99*   trans—*100*

3.5.7-Triphenyl-4H-1.2-diazepin *101* liegt ausschließlich in der monocyclischen Form mit zwei C=N-Bindungen vor, da hier das entsprechende

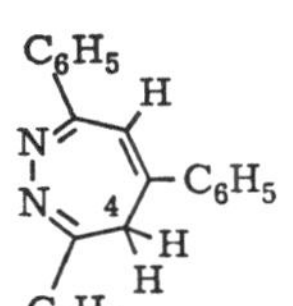

101

Norcaradien *102* eine Azogruppe besäße [129]. Die aus der Temperaturabhängigkeit der Methylensignale in mehreren Lösungsmitteln bestimmte Freie Enthalpie der Aktivierung für das Durchschwingen des Siebenringes liegt bei etwa 17,5 kcal/Mol (90 °C) [129].

3.1.3. 5 H-Dibenzo[c.e]azepine und 11 H-Dibenzo[b.e]azepine

6.7-Dihydro-5 H-dibenzo[c.e]azepine (*24*, X = N—R) und die entsprechenden quartären Ammonium-Salze (*24*, X = $\overset{\oplus}{N}R^1R^2$) liegen ebenso wie die entsprechenden Dibenzo-cycloheptene in chiralen, verdrillten Bootkonformationen *18* bzw. *20* vor. Diesen Konformationsenantiomeren kommt beim Vorliegen großer ortho-Substituenten (*24*, R $\neq$ H, z.B. F, NO$_2$, OCH$_3$) eine beträchtliche Stabilität zu, so daß nach einer Racematspaltung die konformative Umwandlung *18* $\rightleftharpoons$ *20* (*via 19*) in zahlreichen Fällen polarimetrisch verfolgt werden kann [130].

Die Aktivierungsenergien liegen oft zwischen 25 und 40 kcal/Mol. Das nicht durch zusätzliche ortho-Substitutionen (R = H) versteifte N-Benzyl-6.7-dihydro-5 H-dibenzo[c.e]azepin (*24*, R = H, X = N—CH$_2$ —C$_6$H$_5$) ist allerdings ähnlich beweglich ($\Delta G^{\pm}_{63\ °C}$ = 10,1 kcal/Mol) [131] wie der analoge Kohlenwasserstoff (*24*, R = H, X = CH$_2$).

Neuerdings wurde auch für einige 5.6-Dihydro-11 H-dibenzo [b.e] azepin-6-one gezeigt, daß dort im ^{1}H-NMR-Spektrum bei 37 °C die diastereomeren Bootkonformeren, z.B. *103 A* und *103 B* nebeneinander auftreten, d.h. ihre Umwandlung verläuft bei dieser Temperatur, gemessen an der NMR-Zeitskala, relativ langsam [132].

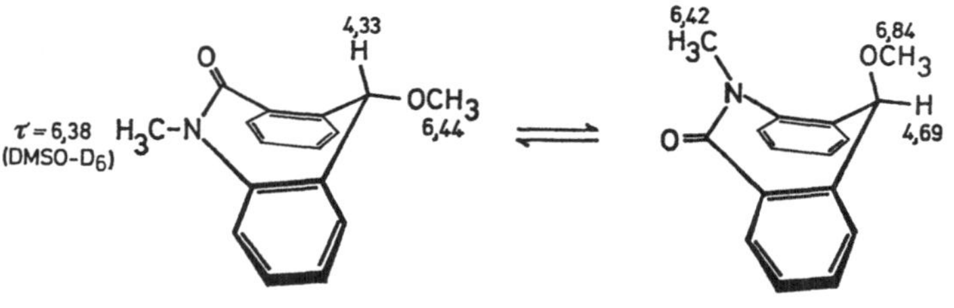

103 A   *103 B*

Die relativen chemischen Verschiebungen der Methin- und Methoxylprotonen (in DMSO—D$_6$ oder CDCl$_3$) entsprechen denjenigen der Systeme *33*, *77* und *79*. Während bei *103* das Konformere *A* überwiegt (K = [103A]:[103B] = 77:23 in CDCl$_3$ bei 37°), kehren sich die Verhältnisse bei der analogen Chlorverbindung (*103*, Cl statt OCH$_3$) um (K = 19:81 unter gleichen Bedingungen) [132].

3.2. Sauerstoff-Heterocyclen

3.2.1. Gesättigte Ringsysteme

Die Konformationsanalyse gesättigter siebengliedriger Sauerstoff-Heterocyclen stößt ebenso wie beim Cycloheptan und dem Perhydro-Azepin-System wegen der großen Beweglichkeit auf Schwierigkeiten. Eine 1970 veröffentlichte Röntgenstrukturanalyse der 1,2;3,4-Di-O-isopropyliden-5-O-chloracetyl-α-D-glucoseptanose *(104)* zeigte, daß der siebengliedrige Ring im Kristall in einer Konformation vorliegt, die zwischen der Sessel- und der Twist-Sessel-Form liegt [133]. Das ^{1}H-NMR-Spektrum in Lösung spricht für eine Sesselkonformation.

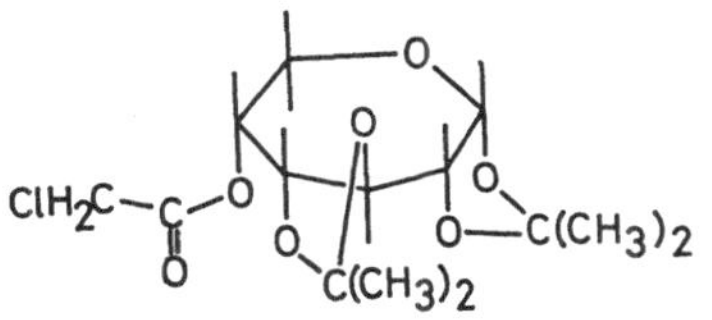

104

3.2.2. Dioxa-cycloheptene (2 H.4 H.7 H-1.3-Dioxepine) und ihre Monobenzo-Derivate

Die ^{1}H-NMR-Spektren der Dioxa-cycloheptene *105* und *106* sowie des Benzo-Derivates *107* zeigen selbst bei tiefen Temperaturen (−120 °C bis −139 °C) keine Aufspaltung (bei *107* lediglich eine Verbreiterung) der Signale von Methylenprotonen und sind demnach recht beweglich. Als einziger O-Heterocyclus dieser Reihe weist das 2.2-Dimethyl-5.6-benzo-2 H.4 H.7 H-1.3-dioxepin (*108*) im Tieftemperatur-Spektrum ein AB-Quartett für die Protonen an den C-Atomen-4 und 7 auf, aus dessen Temperaturabhängigkeit ein $\Delta G^{\pm}_{76\,°C}$-Wert von 9,7 kcal/Mol ermittelt wurde [134]. Da in *108* selbst bei −127 °C die Methylgruppen am C-Atom-2 nicht aufspalten, sondern nur leicht verbreitert erscheinen und somit zumindest eine sehr ähnliche Umgebung haben, wurde für diesen Heterocyclus das Vorliegen einer Twist-Konformation mit C_2-Symmetrie für möglich gehalten [134]. Die Bevorzugung dieser Form bei *108* könnte nach Modellbetrachtungen darauf zurückgeführt werden, daß hier in der Sesselkonformation starke Abstoßungskräfte zwischen einer axialen

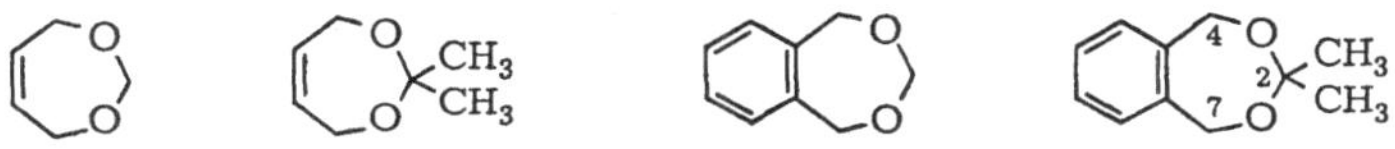

105　　　　*106*　　　　*107*　　　　*108*

Methylgruppe und den axialen Wasserstoffen der C-Atome 4 und 7 zu erwarten wären [134]. Die Umklappschwelle für *108* liegt dabei um ca. 2 kcal/Mol niedriger als beim 5.5-Dimethylbenzocyclohepten *(10)* [33,35].

3.2.3. 5.7-Dihydro-dibenzo[c.e]oxepine

Auch 5.7-Dihydro-dibenzo[c.e]oxepin-Derivate *(16, 17* oder *24,* X = O) sind beweglicher als die entsprechenden Carbocyclen oder Azepin- und Thiepin-Abkömmlinge (s. S. 433). So beträgt die ^{1}H-NMR-spektroskopisch ermittelte Aktivierungsenergie für die Ringbewegung des 5.7-Dihydrodibenzo[c.e]oxepins *(24,* R = H, X = O) nur 9,2 kcal/Mol [135]; der Ersatz aller vier Methylenwasserstoffe durch Methylgruppen erhöht die Umklappschwelle auf $E_a = 13,2$ kcal/Mol [39], während die Einführung einer Nitrogruppe in der freien ortho-Stellung zur Diphenyl-Bindung *(24,* R^1 = H, R^2 = NO_2, X = O) einen kräftigeren Anstieg auf 16,5 kcal/ Mol bedingt [136].

Auch in der Reihe der doppelt verbrückten Biphenyle *17* kommt von den bisher untersuchten Systemen dem Dioxepin *(17,* X = O) die weitaus größte Beweglichkeit zu: Die von Mislow u. Mitarb. [38] optisch aktiv erhaltene Verbindung racemisiert zwischen 10 °C und 23 °C meßbar schnell ($E_a = 20,4$ kcal/Mol, log $k_0 = 12,1$ sec^{-1} in o-Xylol). Die gleiche Substanz wurde später von Oki und Iwamura [137] mit Hilfe der Linienverbreiterung im ^{1}H-NMR-Spektrum untersucht. Die zu 19,9 $\pm$ 0,5 (log $k_0 = 12,0 \pm 0,3$ sec^{-1}) bzw. 21,1 $\pm$ 0,6 kcal/Mol (log $k_0 = 12,2 \pm 0,4$ sec^{-1}) in Hexachlorbutadien bzw. Dimethylsulfoxid $-D_6$ gefundenen Aktivierungsenergien stehen in guter Übereinstimmung mit der polarimetrisch bestimmten [137].

Konformativ beständig werden jedoch auch die Dihydro-oxepine *24* durch Einführung zweier Substituenten in der ortho-Stellung zur Biphenyl-Bindung. Die optisch aktiv vorliegenden Dimethoxy- und Dichlor-Derivate *24* (R = OCH_3 bzw. Cl, X=O) racemisieren erst beim Erhitzen in Lösung ($E_a = 29,9$ kcal/Mol, log $k_0 = 13,4$ sec^{-1} bzw. $E_a = 34,8$ kcal/Mol, log $k_0 = 13,3$ sec^{-1}) [130].

3.2.4. Oxepine und ihre Di- und Tri-benzo-Derivate

Das Vorliegen einer nicht-ebenen, schnell umklappenden Bootkonformation für Oxepine sowie einige einfache Derivate wurde von Vogel und Günther [57] im Rahmen ihrer ^{1}H-NMR- und UV-spektroskopischen Untersuchungen zur Oxepin-Benzoloxid-Valenztautomerie sehr wahrscheinlich gemacht.

So zeigt *109 c* praktisch die gleichen Kopplungskonstanten wie Cycloheptatrien *34* und sein 1.6-Dimethyl-Derivat. Die größere Beweglichkeit der monocyclischen Isomeren *109 a* und *109 b* gegenüber den starren

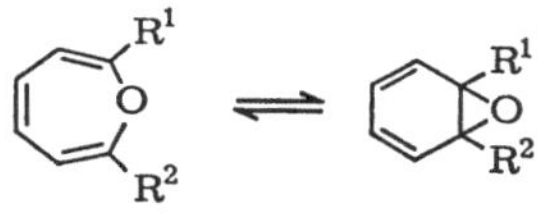

109   110

a: $R^1 = R^2 = H$
b: $R^1 = H, R^2 = CH_3$
c: $R^1 = R^2 = CH_3$

Bicyclen gibt sich durch einen höheren Entropieinhalt der ersteren zu erkennen [138]. Die Bestimmung der Energieparameter für die Ringinversion von *109* steht noch aus.

Dagegen konnten Umklappschwellen für benzologe Oxepine bestimmt werden. *1-Methyl-tribenzo-oxepin-3-carbonsäure 111* liefert mit Brucin ein mutarotierendes Salz, aus dem sich auch die bei Raumtemperatur außerordentlich rasch racemisierende rechtsdrehende Säure ($t_{1/2} =$ 195 sec^{-1} bei 20,5 °C; $\Delta G^{\ddagger}_{20,5 °C} = 20{,}8$ kcal/Mol) gewinnen ließ [139].

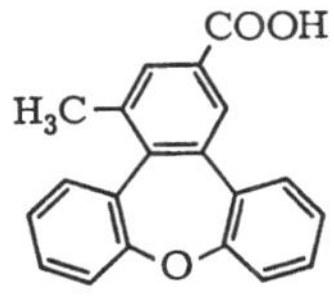

111

Die im Vergleich zu ähnlichen Tribenzo-Systemen verhältnismäßig niedrige Freie Enthalpie der Aktivierung demonstriert wieder die Beweglichkeit von Oxepinen.

Auch das kürzlich von Nogradi, Ollis und Sutherland [89a] untersuchte Dibenzo-oxepin *112* (*61a*; X = O), dessen diastereotope Methylgruppen der Dimethylhydroxy-methyl-Funktion ein temperaturabhängiges ^{1}H-NMR-Spektrum zeigen, weist mit 10,3 kcal/Mol ($T_c = -69$ °C) einen niedrigeren $\Delta G^{\ddagger}$-Wert auf als analoge Ringsysteme (X = CR$_2$, S, N—R).

3.3. Schwefel-Heterocyclen

3.3.1. Gesättigte Ringsysteme

Untersuchungen zur konformativen Beweglichkeit von gesättigten Thiepan-Derivaten verdanken wir vor allem der Freiburger [140,141] sowie neuerdings auch einer anderen Gruppe [142]. Dabei war für Siebenringe durch den Einbau von Disulfid- und vor allem Trisulfid-

Gruppen wegen ihrer hohen Torsionsbarriere und der Bevorzugung eines Torsionswinkels von etwa 90° [143] eine Erhöhung der sehr niedrigen Pseudorotationsbarriere des Cycloheptans [25-27] (ca. 2—3 kcal/Mol) zu erwarten.

Tatsächlich läßt sich auch beim *1.2.3-Trithiepan 113* [140] unterhalb —130 °C ($\Delta G^{\neq} = 6$—7 kcal/Mol) eine intramolekulare Bewegung (Ringinversion?) verlangsamen. Beim 3.3.7.7-Tetradeutero-5.5-dimethyl-1.2-dithiepan *(114)* [141] sind in den Tieftemperatur-Spektren (zwei AB-Systeme für die Methylenprotonen) zwei Konformere *A* und *B* im Besetzungsverhältnis 69:31 (in CS_2) zu erkennen und beim kontinuierlichen Abkühlen können auch zwei verschiedene Umwandlungsprozesse ($\Delta G^{\neq}_{-73 °C} = 10{,}8$ kcal/Mol; $\Delta G^{\neq}_{-110 °C} = 8{,}0$ kcal/Mol) nacheinander eingefroren werden. Aufgrund der zahlreichen Umwandlungsmöglichkeiten des gesättigten Siebenringes [25,141] (s. S. 383—386) ist selbst aus diesen Daten eine zuverlässige Konformationsangabe für *A* und *B* sowie eine Zuordnung der beobachteten Prozesse nicht ohne weiteres möglich [141]. Da es sich jedoch bei einem dieser verlangsamten Vorgänge sicher um einen Pseudorotationsschritt handeln wird, ist damit gezeigt, daß die Einführung von S-S-Bindungen zu einer merklichen Erhöhung der Pseudorotations-Barriere des gesättigten Siebenringes führt [25,141].

Während für das Disulfid *115* ebenfalls ein $\Delta G^{\neq}_{-89 °C}$-Wert von 9 kcal/Mol ermittelt wurde, sind die Ringsysteme *116* und *117*, die keine S-S-Bindungen enthalten, noch wesentlich beweglicher ($\Delta G^{\neq} < 8$ kcal/Mol) [141].

| 114 | 115 | 116 | 117 | 118 |

Andererseits beobachtet man beim 1.2.3.5.6-Pentathiepan (Lenthionin) *118* eine merkliche Versteifung des Ringgerüstes [142]: Das ^{1}H-NMR-Spektrum von *118* zeigt bei 30 °C nur ein einziges Singulett für die Methylenprotonen, das bei $T_c = -60$ °C aufspaltet und bei —90 °C in zwei scharfe Singulett-Signale übergeht. Zur Deutung wird angenommen [142], daß es sich bei dem einfrierbaren Vorgang um die Ringinversion ($E_a = 12{,}9 \pm 0{,}4$ kcal/Mol) zwischen den Konformationen *118 A* und *118 B*, die nicht äquivalente Paare von Methylengruppen besitzen, handelt. Die Annahme einer derartigen Sesselform wird hier durch die Röntgenstrukturanalyse des kristallinen *118* gestützt [144].

118 A   *118 B*

Das Nichtauffinden einer geminalen Kopplung zwischen den Methylenprotonen wird von den Autoren auf einen zweiten schnellen Pseudorotationsvorgang, der einen Austausch geminaler Wasserstoffe bewirkt, zurückgeführt [142]. Eine solche Erklärung ist grundsätzlich möglich, da auch bei langsamem Ablauf einzelner Pseudorotationsschritte des gesättigten Siebenringes andere Teilschritte des Pseudorotationscyclus noch sehr schnell velaufen können.

Das offenbar beweglichere Tetrathiepan *119* zeigt zwischen Raumtemperatur und $-90\,°C$ ein unverändertes Spektrum: Zwei scharfe Resonanzlinien im Verhältnis 2:1 wurden den strukturell verschiedenen Gruppen von Methylenprotonen in der Sesselkonformation *119 A* zugeschrieben [142].

119   *119 A*

3.3.2. Di- und Tri-thiacycloheptene (2 H.4 H.7 H-1.3-Dithiepine, 3 H.6 H.7 H-1.2-Dithiepine, 5 H.6 H.7 H-1.4-Dithiepine und 4 H.7 H-1.2.3-Trithiepine) und ihre Monobenzo-Derivate

Die energieärmste Konformation der Dithiepine *120* und *121* ist sehr wahrscheinlich ebenfalls die Sesselform. Dies folgt aus dem Auftreten eines Dubletts für die geminalen Methylgruppen und eines AB-Quartetts für die Methylenprotonen. Auch hier sind die Freien Enthalpien der Aktivierung verglichen mit verwandten Monocyclen relativ hoch ($\Delta G^{\ddagger}_{-100\,°C}$ = 8,5 bzw. $\Delta G^{\ddagger}_{105\,°C}$ = 8,2 kcal/Mol) [134].

120   *121*   *122*

Bemerkenswerterweise erscheinen im Tieftemperatur-Spektrum von
121 zusätzliche Signale für die Substituenten an den C-Atomen 2 und 4
bzw. 7, d.h. hier liegt im Gleichgewicht ein weiteres Konformeres, vermutlich in der flexiblen Wannenform (Pseudorotationsmöglichkeit zwischen normaler Wanne und Twistwanne *8b* und *8c*) vor. Die Sesselform
ist dabei um $\Delta G_{-120°} = 0,2$ kcal/Mol günstiger [134]. Wie zu erwarten,
läßt sich auch bei dem Trisulfid *122* ein Umwandlungsvorgang einfrieren
($\Delta G^{\pm}_{-90 \, °C} = 8,9$ kcal/Mol) [140].

Ähnlich wie bei den Benzocycloheptenen [33-35] (s. S. 390) liegen auch
umfangreiche Studien zur Beweglichkeit von Benzodithia- und -trithia-
cycloheptenen aus der „Freiburger Gruppe" [29,32,134,141] vor. Diese Untersuchungen sind deshalb von besonderem Interesse, da einmal in den
Tieftemperatur-Spektren dieser Verbindungsklasse oft zwei Konformere
nebeneinander sichtbar sind und außerdem in mehreren Fällen die Energieschwelle verschiedener Umwandlungsprozesse (Version (*V*) und Pseudorotation (*P*)) an ein und demselben Molekül bestimmt werden können.
Dies gelang zuerst am Beispiel der Verbindungen *123—125* [29]:

	123	*124*	*125*
Besetzungsverhältnis (CS$_2$) S: (T bzw. W)	45:55	85:15	70:30
Temperatur	25 °C	25 °C	−35 °C
$\Delta G^{\pm}_c$ (*V*) kcal/Mol	19,8	17,4	13,5
Temperatur T_c	114 °C	83 °C	0 °C
E_a kcal/Mol	20	—	—
$\Delta G^{\pm}_c$ (*P*) kcal/Mol	11,5	∼10	10,4
Temperatur T_c	−45 °C	∼ −80 °C	−60 °C
E_a	14	—	—

Lösungsmittel: C$_4$Cl$_6$ oberhalb von + 25 °C; CS$_2$ unterhalb von 25 °C.

So zeigt *123* oberhalb 165 °C jeweils scharfe Singulettsignale für die
Phenyl-, Methylen- und Methylprotonen. Bei Raumtemperatur findet
man dagegen im Bereich der Phenyl- und Methylprotonen jeweils zwei
Signale im Intensitätsverhältnis 45:55, im Bereich der Methylenprotonen
ein AB-Quartett und ein zusätzliches scharfes Signal, die ebenfalls im
Verhältnis 45:55 stehen. Unterhalb −45 °C wird schließlich auch die

zuletzt genannte Resonanz zu einem AB-Quartett aufgespalten. Diese Befunde können folgendermaßen gedeutet werden: Bei hohen Temperaturen [$T > 114\,°C$] wandeln sich alle möglichen Konformeren sehr rasch ineinander um. Bei Raumtemperatur ist dagegen schon die Version (wahrscheinlich der geschwindigkeitsbestimmende Schritt der Sesselinversion) verlangsamt und man erhält die Signale zweier verschiedener Konformerer. Das AB-Quartett und die noch nicht aufgespaltene Resonanz im Bereich der Methylenprotonen können der „starren" Sesselform und der „flexiblen" Wannen- oder Twist-Form zugeordnet werden. Bei *123* überwiegt im Gleichgewicht sogar das flexible Konformere ($\Delta G = 0{,}12$ kcal/Mol bei Raumtemperatur). Unterhalb $-45\,°C$ läßt sich dann auch die Pseudorotation der flexiblen Form einfrieren, so daß auch deren Methylenprotonen ein AB-Quartett im NMR-Spektrum liefern. Bei *124* und *125* ergaben sich ähnliche Verhältnisse; dort ist allerdings die Sesselkonformation am energieärmsten.

Im Disulfid *126* [141] liegen ebenfalls zwei Konformere mit nahezu gleichem Besetzungsverhältnis vor. Ähnlich wie bei den Benzocycloheptenen (s. S. 390) konnte dabei aufgrund der relativ kleinen chemischen Verschiebungsdifferenz der geminalen Methylgruppen ($\Delta v = 0{,}31$ ppm) dem starren Konformeren die Sesselform zugeordnet werden. Für die ausgeschlossene Wannenform wäre auch hier ein sehr viel größerer Unterschied zu erwarten gewesen.

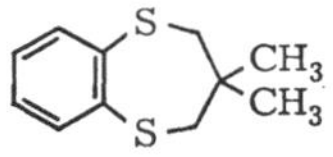

126

Besetzungsverhältnis:

S:T (in CS_2)	49:51
$\Delta G^{\neq}$ (*V*) kcal/Mol	12,5
Temperatur	$-28\,°C$
$\Delta G^{\neq}$ (*P*) kcal/Mol	8,4
Temperatur	$-90\,°C$

Der Befund, daß das Signal der Methylgruppen des erst bei tieferen Temperaturen einfrierbaren „flexiblen" Konformeren bis $-112\,°C$ nicht aufspaltet (lediglich ein AB-System für die Methylenprotonen unterhalb von $-90\,°C$) ist ein überzeugender Hinweis dafür, daß hier das flexible Konformere die Twistkonformation (T) mit C_2-Symmetrie *(8c)* besitzt. Demnach ergibt sich aus dem Aufspaltungsbild der Methylenprotonenresonanzen, daß der langsamere der beiden Prozesse sowohl die Umwandlung der Sessel- und der Twist-Form ineinander als auch die Ringinversion des Sessels bewirkt, während der schnellere Vorgang der Inversion der Twist-Form zuzuschreiben ist. Aufgrund des auf S. 391 angegebenen Umwandlungsschemas lassen sich alle Ergebnisse so deuten, daß ein Versionsvorgang (z. B. S ⇌ T) der geschwindigkeitsbestimmende Teilschritt der Sesselinversion ist, während das Umklappen der Twist-Form durch Pseudorotation erfolgt. Die Energiebarrieren der beiden Prozesse unterscheiden sich hier um $\Delta G^{\neq} (V) - \Delta G^{\neq} (P) \approx 4$ kcal/Mol.

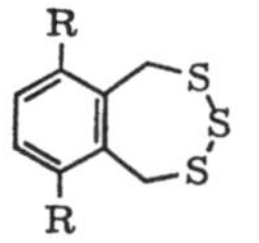

127

a: R = H
b: R = OCH$_3$
c: R = CH$_3$
d: R = C$_6$H$_5$

Für das starre Konformere in der Reihe der Benzo-trithiepine *127* wurde ebenfalls die Sesselform, für das flexible dagegen — vor allem wegen des bevorzugten Torsionswinkels von 90° der S-S-Bindungen — die Wannenform angenommen [141]. Bei *127b—127d* erfordert der langsame Prozeß einen beträchtlichen Energieaufwand, zwischen 19,5 und 21,2 kcal/Mol, während der schnellere nur ca. 10—11 kcal/Mol benötigt. Interessanterweise kristallisierten die Trisulfide *127c* und *127d* in der mutmaßlichen, flexiblen Wannenform; *127b* dagegen in der Sesselform (Reinheitsgrad >95%). Damit besteht die Möglichkeit, durch Röntgenstrukturanalyse die Konformation im Kristall zu bestimmen. Außerdem wurde die Einstellung der entsprechenden Konformerengleichgewichte durch Auflösung der Kristalle bei ca. —40 °C und Verfolgung der Zeitabhängigkeit der NMR-Spektren gemessen [141].

3.3.3. 5.7-Dihydro-dibenzo[c.e]thiepine und 10.11-Dihydro-dibenzo-[b.f]thiepine

Von den bislang untersuchten verbrückten Biphenylen *16*, *17* oder *24* mit siebengliedrigem Ring kommt den Thiepinen und ihren S.S-Dioxid-Derivaten (X = S oder SO$_2$) die größte konformative Stabilität zu [38, 130]. So steigt die NMR-spektroskopisch bestimmte Energieschwelle für den Umklappvorgang *18 → 20* beim Übergang vom Oxepin (*16*, X = O; $E_a = 9,2$ kcal/Mol; E (ber.) = 9 kcal/Mol) zum Thiepin (*16*, X = S) auf

$E_a = 16,1$ kcal/Mol (E (ber.) $= 17$ kcal/Mol) an [38,39,135], Das analoge Thiepin-S.S-dioxid (16, $X = SO_2$) zeigt mit $G^{\ddagger}_{87,5\,°C} = 18,2$ kcal/Mol sogar eine noch höhere Umklappbarriere [39].

Das gleiche Phänomen beobachtet man auch in der Reihe der doppelt verbrückten Systeme 17, wie aus der folgenden Tabelle hervorgeht [38].

Aktivierungsparameter für den Umklappvorgang doppelt verbrückter Biphenyle 17

17	X=O	X¹=O, X²=S	X=CO	X=S
E_a (kcal/Mol) *)	20,4	30,6	31,2	35,0
log k_0 (sec⁻¹)	12,1	13,5	14,5	12,2
E (ber.) (kcal/Mol)	22		33	37

*) Polarimetrisch in o-Xylol als Lösungsmittel bestimmt.

Dabei ergibt sich, daß die konformative Stabilität dieser Biphenyle mit steigendem Torsionswinkel ϕ im Grundzustand zunimmt (ϕ ist der Winkel, den die Ebenen der beiden Benzolringe miteinander einschließen und stellt somit ein Maß für die Verdrillung des Biphenyl-Skeletts um die 1.1′-Bindung dar). Dieser Zusammenhang ist auch leicht verständlich: Nimmt man an, daß im Übergangszustand 19 der Ringbewegung der Torsionswinkel $\phi = 0$ wird, so folgt, daß die Veränderung dieses Winkels beim Übergang vom Grund- zum Übergangszustand (18 → 19) um so größer sein muß, je größer ϕ bereits im Grundzustand 18 bzw. 20 ist.

Nachstehend sind die berechneten Torsionswinkel ϕ für einige einfach verbrückte Biphenyle 16 angegeben [38].

16	X = O	X = N−CH₃	X = CH₂	X = CO	X = S
ϕ	44,1°	45,8°	50,6°	52,4°	56,6°

Diese Veränderungen von ϕ schlagen sich auch in den NMR- und UV-Spektren nieder [38,130], wobei eine stärkere Verdrillung im allgemeinen zu einer erhöhten Abschirmung der Methylenprotonen (bei 24 mit $R = CH_3$ auch der Methylprotonen) im NMR und zu einer hypsochromen Verschiebung der Konjugationsbande im UV führt [38,130].

Diese Spektren zeigen außerdem, daß der Torsionswinkel ϕ bei gleichem Ringglied X in der Reihenfolge 17 < 16 < 24 ($R = CH_3$) zunimmt [38,130]. Bemerkenswert ist die gute Übereinstimmung der von Mislow [38] berechneten Aktivierungsenergien für die Ringinversion mit den NMR-spektroskopisch oder polarimetrisch bestimmten Werten [38].

Die absolute Konfiguration mehrerer optisch aktiv vorliegender Dihydro-oxepine und -thiepine konnte mit Hilfe von ORD- und CD-Kurven bestimmt werden [38,130].

NMR-spektroskopische Untersuchungen an 10.11-Dihydro-dibenzo-[b.f]thiepinen *128*, bei denen die Benzolringe nicht direkt miteinander verbunden sind, ergaben eine sehr viel höhere Beweglichkeit dieser Verbindungsklasse. ($\Delta G^{\pm}_{-80\,°C} = 9,3$ kcal/Mol für X = S und $\Delta G^{\pm}_{-44\,°C} = 10,7$ kcal/Mol für X = SO_2) [145].

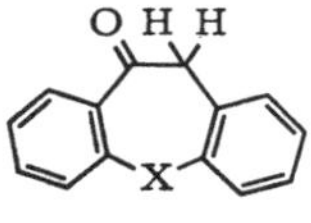

128

X = S, SO_2

3.3.4. Thiepine, Thiepin-S.S-dioxide, ihre Di- und Tri-benzo-Derivate sowie verwandte Heterocyclen

Die Röntgenstrukturanalyse des von Mock [146] erstmals dargestellten Thiepin-S.S-dioxids *(129a)* ergab eine Bootkonformation (vgl. *34*) mit Strukturwinkeln von $\alpha = 44,6°$ und $\beta = 22,8°$ [147]. Die UV- und ^{1}H-NMR-Spektren von *129* sprechen dafür, daß dieser Heterocyclus auch in Lösung als flaches Boot vorliegt [146,148]. Kürzlich wurde die Energieschwelle des Boot-Inversion des 3-Isopropyl-6-methyl-thiepin-S.S-dioxids *(129b)* mit Hilfe der Temperaturabhängigkeit der Signale für die diastereotopen Methylgruppen $(CH_3)_A$ und $(CH_3)_B$ zu $\Delta G^{\pm}_{150\,°C} = 6,4$ kcal/Mol bestimmt [149]. Dieser Wert ist demjenigen des Cycloheptatriens *(34)* sehr ähnlich, obwohl die längeren C—S-Bindungen und der kleinere C—S—C-Bindungswinkel (1,72 Å und 103,3° in *129a*) zu einer Erschwerung der Ringinversion gegenüber *34* hätten führen können. Zur Erklärung werden Konjugationseffekte diskutiert [149].

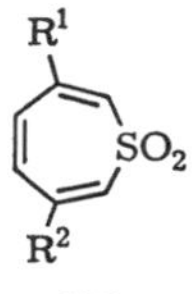

R^1

SO_2

R^2

129

a: $R^1 = H$, $R^2 = H$
b: $R^1 = CH_3$, $R^2 = CH(CH_3)_A(CH_3)_B$

Genau wie bei Cycloheptatrienen so bewirkt auch die Annellierung von drei Benzolringen an das Thiepin-S.S-dioxid-Gerüst eine drastische Erhöhung der Energiebarriere für die Ringbewegung; daraus ergibt sich die Möglichkeit der Spaltung chiraler Derivate in bei Raumtemperatur beständige Enantiomere. Dies gelang 1967 bei der Tribenzo-thiepin-S.S-dioxid-2-carbonsäure *(130)* [150)], die erst beim Erhitzen in Lösung zwischen 80 °C und 110 °C langsam racemisiert ($\Delta G^{\ddagger}_{80\,°C} = 29,5$ kcal/Mol; $\Delta S^{\ddagger}_{80\,°C} = +5$ (± 5) e.u.; $E_a = 32 \pm 1,5$ kcal/Mol in Diäthylenglykoldimethyläther).

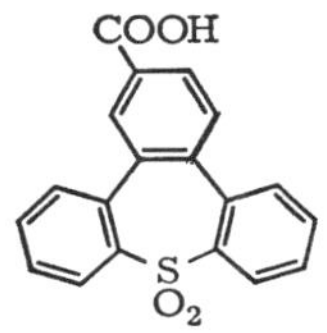

COOH

130

Auch die Gleichgewichtseinstellung zwischen diastereomeren Dibenzo[b.f]1.4-thiazepin-S-oxiden der Struktur *131 A* und *131 B* (mit äquatorialem und axialem Sauerstoff an der SO-Gruppe) durch Umklappen des Siebenringes erfolgt oberhalb von Raumtemperatur meßbar langsam (E_a *(131 A → 131 B)* = 21 kcal/Mol, log $k_0 = 10,13$ sec^{-1}; E_a *(131 B → 131 A)* = 20,3 kcal/Mol, log $k_0 = 10,16$ sec^{-1}) [151)].

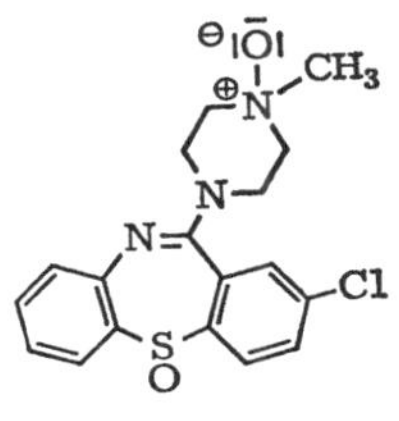

CH₃

Cl

131

Kürzlich wurde die Geschwindigkeit der Ringinversion eines Dibenzothiepin-Derivates *112 (61 a;* X = S) ¹H-NMR-spektroskopisch ermittelt. Die zu 17,7 kcal/Mol bei 51 °C bestimmte Freie Enthalpie der Aktivierung liegt — ebenso wie bei den Dihydrothiepinen *16* und *17* — höher als beim analogen Dibenzooxepin *112 (61 a;* X = O) [89a)].

Abschließend sei noch einschränkend [156,157)] darauf hingewiesen, daß die in dieser Übersicht besprochenen und häufig miteinander verglichenen Aktivierungsparameter für Ringbewegungen oft in unterschiedlichen Lösungsmitteln erhalten wurden.

Bis jetzt scheinen allerdings keine dramatischen Beeinflussungen dieser Daten bekannt zu sein, solange die in der Protonenresonanz-Spektroskopie und Polarimetrie gebräuchlichen Solvenzien verwendet wurden. Selbst bei Stickstoff-Heterocyclen, für die in einigen Fällen das Lösungsmittel variiert wurde (*88* [121], *94* [123], *101* [129]) ergaben sich nur kleine oder überhaupt keine Unterschiede in den gemessenen Werten [121,123,129].

Größere Effekte sind wohl in erster Linie bei Lösungen von Heterocyclen in sehr polaren Medien zu erwarten, wenn außer der Ringbewegung auch noch andere Vorgänge (Protonierungen etc.) zu berücksichtigen sind (z.B. *101* in Trifluoressigsäure [129]).

4. Schluß

Die Übersicht sollte zeigen, daß es heute in vielen Fällen möglich ist, zumindest qualitative oder halbquantitative Aussagen über bevorzugte Konformationen bei siebengliedrigen Ringsystemen zu machen. Zwar liegen nur in vergleichsweise wenigen Fällen Röntgenstruktur- oder Elektronenbeugungs-Analysen vor, so daß die möglichst exakte Topographie vieler Konformationen nicht bekannt ist. In ungünstigen Fällen, die sich etwa für eine detaillierte (an Hand von chemischen Verschiebungen und vor allem von Kopplungsparametern) NMR-Analyse nicht eignen, ist man letztlich doch auf Modellbetrachtungen und qualitative Schlüsse aus den (NMR-, UV- etc.) Spektren angewiesen. Hier bleibt für die Zukunft sicher noch viel zu tun.

Andererseits können durch eine sinnvolle Kombination physikalischer Methoden (in erster Linie NMR-Spektroskopie und Polarimetrie) die Gleichgewichtslagen zwischen Konformeren und die Aktivierungsparameter für Ringbewegungen in vielen Fällen ermittelt werden.

Das Wissen um die Konformation von Molekülen und um die Energiegrößen für konformative Prozesse ist auch bei den siebengliedrigen Ringen nicht nur von bloßem Erkenntniswert, sondern bildet darüber hinaus die Grundlage zur Deutung anderer Eigenschaften dieser Verbindungen, wie an zwei Beispielen aufgezeigt sei.

Da ist einmal die Möglichkeit zur Erklärung konformativ bedingter Reaktivitätsunterschiede zu nennen [45,90,91,92,117,152] (s.S. 416ff.).

Außerdem besitzen zahlreiche Verbindungen mit siebengliedrigem Ring in der Medizin eine ständig zunehmende Bedeutung als *Arzneimittel* [153]. Hier ist nun interessanterweise die Hypothese aufgestellt worden, daß einer der wichtigen Faktoren, die etwa die mehr neuroleptische oder thymoleptische Wirkungsweise eines Psychopharmakons bestimmen, die Konformation des zentralen Siebenringes ist [153,154]. Ähnliches soll auch für analoge Substanzen mit Sechsring-Struktur gelten. Danach nimmt die thymoleptische Potenz mit zunehmendem Verdrillungsgrad der zentralen Ringe Z bei den tricyclischen Psychopharmaka der allgemeinen Struktur *132* zu [153,154].

132

Somit ist zu hoffen, daß die hier geschilderten Ergebnisse eines Tages auch über den Rahmen der ,,Dynamischen Stereochemie'' hinaus von Interesse sein werden.

Danksagung. — Die hier zitierten eigenen Arbeiten wurden von der Deutschen Forschungsgemeinschaft sowie dem Fonds der Chemischen Industrie nachhaltig gefördert, denen auch an dieser Stelle gedankt sei. Herrn Priv.-Doz. Dr. A. Mannschreck möchte ich für seine Zusammenarbeit sowie für wertvolle Diskussionen danken. Ihm sowie den Herren Doz. Dr. S. Kabuß und Dr. S. Schütz gilt auch mein Dank für eine kritische Durchsicht des Manuskriptes. Für die Anfertigung von Formelzeichnungen danke ich Herrn Dipl.-Chem. G. Schmidt.

Literatur

[1] Eliel, E. L., Allinger, N. L., Angyal, S. J., Morrison, G. A.: Conformational Analysis. New York: Interscience Publishers 1965.

[2] Hanack, M.: Conformation Theory. New York: Academic Press 1965.

[2a] Zur Definition dieses Begriffes s. auch Nach. Chem. Techn. *17*, 443 (1969).

[3] Barton, D. H. R.: Experientia *6*, 316 (1950).

[4] Siehe Lehrbücher der Stereochemie, z.B.
   [a] Eliel, E. L.: Stereochemie der Kohlenstoffverbindungen. Weinheim/Bergstraße: Verlag Chemie GmbH 1966.
   [b] Mislow, K.: Einführung in die Stereochemie. Weinheim/Bergstraße: Verlag Chemie GmbH 1967.

[5] Pople, J. A., Schneider, W. G., Bernstein, H. J.: High Resolution Nuclear Magnetic Resonance. New York: McGraw Hill Book Company 1959.

[6] Roberts, J. D.: Nuclear Magnetic Resonance, Applications to Organic Chemistry. New York: McGraw Hill Book Company 1959.

[7] Emsley, J. W., Feeney, J., Sutcliffe, L. H.: High Resolution Nuclear Magnetic Resonance Spectroscopy, Vol. 1. Oxford: Pergamon Press 1965.

[8] Bovey, F. A.: Nuclear Magnetic Resonance Spectroscopy, S. 183. New York: Academic Press 1969.

[9] Loewenstein, A., Connor, T. M.: Ber. Bunsenges. Physik. Chem. *67*, 280 (1963).

[10] Delpuech, J. J.: Bull. Soc. Chim. France *1964*, 2697.

[11] Reeves, L. W.: In: Advances in Physical Organic Chemistry, Band 3, S. 187; Gold, V., Hrsg. New York: Academic Press 1965.

[12] Johnson, C. S.: In: Advances in Magnetic Resonance, Vol. 1, S. 33; Waugh, J. S., Hrsg. New York: Academic Press 1965.

[13] Anderson, J. E.: Quart. Rev. (London) *19*, 426 (1965).

[14] Binsch, G.: Topics in Stereochemistry, Band 3, S. 97; Allinger, N. L., Eliel, E. L., Hrsg. New York: Interscience Publishers 1968.

[15] Tochtermann, W., Schnabel, G., Mannschreck, A.: Liebigs Ann. Chem. *705*, 169 (1967).

[16] Bei Cycloheptatrien-Derivaten ist a priori nur eine koplanare Anordnung aller Atome im Siebenring oder die nichtebene Boot- bzw. Wannenform zu diskutieren. Eine Sesselform, die trans-konfigurierte Doppelbindungen erfordern würde, ist hier aus Spannungsgründen nicht möglich.

[16a] Zur Definition s. [4b], und zwar S. 68, sowie Mislow, K., Raban, M.: Topics in Stereochemistry, Band 1, S. 1; Allinger, N. L., Eliel, E. L., Hrsg. New York: Interscience Publishers 1967.

[17] Literaturzusammenstellung bei [14].

[18] Frost, A. A., Pearson, R. G.: Kinetics and Mechanisms, S. 24, 99. New York: J. Wiley and Sons, Inc. 1961.

[19] Zur Anwendung der Doppelresonanz- und Spin-Echo-Methoden auf derartige Probleme s. Forsén, S., Hoffman, R. A.: J. Chem. Phys. *39*, 2892 (1963). — Anet, F. A. L., Bourn, A. J. R.: J. Am. Chem. Soc. *89*, 760 (1967). — Allerhand, A., Gutowsky, H. S.: J. Am. Chem. Soc. *87*, 4092 (1965). — Abramson, K. H., Inglefield, P. T., Krakower, E., Reeves, L. W.: Can. J. Chem. *44*, 1685 (1966).

[20] Allerhand, A., Gutowsky, H. S., Jonas, J., Meinzer, R. A.: J. Am. Chem. Soc. *88*, 3185 (1966), dort weitere Lit.

[21] Schmid, H. G., Friebolin, H., Kabuß, S., Mecke, R.: Spectrochim. Acta *22*, 623, (1966).

[22] Jaeschke, A., Muensch, H., Schmid, H. G., Friebolin, H., Mannschreck, A.: J. Mol. Spectry. *31*, 14 (1969).

[23] Gutowsky, H. S., Holm, C. H.: J. Chem. Phys. *25*, 1228 (1956).

[24] Mannschreck, A., Mattheus, A., Rissmann, G.: J. Mol. Spectry. *23*, 15 (1967).

[24a] Weitere Beispiele: Ollis, W. D., Sutherland, I. O.: Chem. Commun. *1966*, 402. — Downing, A. P., Ollis, W. D., Sutherland, I. O.: J. Chem. Soc. B *1970*, 24. — Gutowsky, H. S., Jonas, J., Siddall, T. H., III: J. Am. Chem. Soc. *89*, 4300 (1967); sowie [137].

[25] Hendrickson, J. B.: J. Am. Chem. Soc. *83*, 4537 (1961); *84*, 3355 (1962).

[26a] Siehe dazu auch Pauncz, R., Ginsburg, D.: Tetrahedron *9*, 40 (1960).

[b] Allinger, N. L., Szkrybalo, W.: J. Org. Chem. *27*, 722 (1962).

[c] Bixon, M., Lifson, S.: Tetrahedron *23*, 769 (1967).

[27] Zur Definition des Begriffes Pseudorotation siehe Kilpatrick, J. E., Pitzer, K. S., Spitzer, R.: J. Am. Chem. Soc. *69*, 2483 (1947) sowie Lit. [1], S. 201—208, [4a] S. 304; [13]. Dieser Begriff wird auch noch für einen anderen Vorgang, nämlich die intramolekularen Bewegungen von trigonalen bipyramidalen Strukturen bei fünfbindigen Atomen verwendet; s. dazu Mutterties, E. L., Schunn, R. A.: Quart. Rev. (London) *20*, 245 (1966). — Westheimer, F. H.: Accounts Chem. Res. *1*, 70 (1968).

[28] Roberts, J. D.: Chem. Brit. *1966*, 529.

[29] Kabuß, S., Lüttringhaus, A., Friebolin, H., Schmid, H. G., Mecke, R.: Tetrahedron Letters *1966*, 719.

[30] Lambert, J. B., Roberts, J. D.: J. Am. Chem. Soc. *87*, 3884 (1965).

31) Knorr, R., Ganter, C., Roberts, J. D.: Angew. Chem. 79, 577 (1967); Angew. Chem. Intern. Ed. Engl. 6, 556 (1967).
31a) S. dazu Lit. 1), S. 206—210. — Mann, G., Mühlstädt, M., Müller, R., Kern, E., Hadeball, W.: Tetrahedron 24, 6941 (1968). — Baumann, H., Möhrle, H.: Tetrahedron 25, 135 (1969); dort weitere Literatur.
31b) Eistert, B., Haupter, F., Schank, K.: Liebigs Ann. Chem. 665, 55 (1963). — Schank, K., Eistert, B.: Chem. Ber. 99, 1414 (1966).
32) Friebolin, H., Mecke, R., Kabuß, S., Lüttringhaus, A.: Tetrahedron Letters 1964, 1929.
33) Kabuß, S., Friebolin, H., Schmid, H.: Tetrahedron Letters 1965, 469.
34) — Schmid, H. G., Friebolin, H., Faißt, W.: Org. Magnetic Resonance 1, 451 (1969) (XIII. Mitteil.).
35) — — — — Org. Magnetic Resonance 2, 19 (1970). (XIV. Mitteil.).
36) Grunwald, E., Price, E.: J. Am. Chem. Soc. 87, 3139 (1965).
37) Mislow, K.: Angew. Chem. 70, 683 (1958).
38) — Hyden, S., Schaefer, H.: J. Am. Chem. Soc. 84, 1449 (1962). — Mislow, K., Glass, M. A.W., Hopps, H. B., Simon, E., Wahl, G. H., Jr.: J. Am. Chem. Soc. 86, 1710 (1964) und dort zitierte Literatur.
39) Sutherland, I. O., Ramsay, M. V. J.: Tetrahedron 21, 3401 (1965).
40) Iffland, D. C., Siegel, H.: J. Am. Chem. Soc. 80, 1947 (1958).
41) Eyring, H.: J. Chem. Phys. 3, 107 (1935).
42) Ebnöther, A., Jucker, E., Stoll, A.: Helv. Chim. Acta 48, 1237 (1965).
43) Zusammenstellung und Literatur bei Mislow, K., Glass, M. A. W., O'Brien, R. E., Rutkin, P., Steinberg, D. H., Weiss, J., Djerassi, C.: J. Am. Chem. Soc. 84, 1455 (1962).
44) Bunnenberg, E., Djerassi, C., Mislow, K., Moscowitz, A.: J. Am. Chem. Soc. 84, 2823 (1962). — Mislow, K., Bunnenberg, E., Records, R., Wellman, K., Djerassi, C.: J. Am. Chem. Soc. 85, 1342 (1963).
45) Lansbury, P. T.: Accounts Chem. Res. 2, 210 (1969); dort weitere Literatur.
46) — Bieron, J. F., Klein, M.: J. Am. Chem. Soc. 88, 1477 (1966).
47) — — J. Am. Chem. Soc. 86, 2524 (1964).
48) Biffin, M. E. C., Crombie, L., Connor, T. M., Elvidge, J. A.: J. Chem. Soc. B 1967, 841.
49) Lansbury, P. T., Bieron, J. F., Lacher, A. J.: J. Am. Chem. Soc. 88, 1482 (1966).
50) — Lacher, A. J., Saeva, F. D.: J. Am. Chem. Soc. 89, 4361 (1967).
51) Colson, J. G., Lansbury, P. T., Saeva, F. D.: J. Am. Chem. Soc. 89, 4987 (1967).
52) Übersicht zum Cycloheptatrien-Norcaradien-Problem: Maier, G.: Angew. Chem. 79, 446 (1967); Angew. Chem. Intern. Ed. Engl. 6, 402 (1967).
53) Siehe Doering, W. v. E., Knox, L. H.: J. Am. Chem. Soc. 76, 3203 (1954). — Doering, W. v. E., Laber, G., Vonderwahl, R., Chamberlain, N. F., Williams, R. B.: J. Am. Chem. Soc. 78, 5448 (1956).
54a) Abel, E. W., Bennett, W. A., Wilkinson, G.: Proc. Chem. Soc. (London) 1958, 152.
54b) Evans, M. V., Lord, R. C.: J. Am. Chem. Soc. 82, 1876 (1960).
54c) La Lau, C., De Ruyter, H.: Spectrochim. Acta 19, 1559 (1963).
55) Corey, E. J., Burke, H. J., Remers, W. A.: J. Am. Chem. Soc. 77, 4941 (1955).
56) Conrow, K.: J. Am. Chem. Soc. 83, 2958 (1961).
57) Übersicht zur Benzoloxid-Oxepin-Valenztautomerie: Vogel, E., Günther, H.: Angew. Chem. 79, 429 (1967); Angew. Chem. Intern. Ed. Engl. 6, 385 (1967).
58) Ciganek, E.: J. Am. Chem. Soc. 87, 1149 (1965).
58a) Berson, J. A., Hartter, D. R., Klinger, H., Grubb, P. W.: J. Org. Chem. 33, 1669 (1968).

[59] Görlitz, M., Günther, H.: Tetrahedron 25, 4467 (1969). — Günther, H.: Privatmitteilung.

[60] Ciganek, E.: J. Am. Chem. Soc. 87, 652 (1965).

[61] Mukai, T., Kubota, H., Toda, T.: Tetrahedron Letters 1967, 3581.

[62] Schönleber, D.: Angew. Chem. 81, 83 (1969); Angew. Chem. Intern. Ed. Engl. 8, 76 (1969).

[63] Jones, M., Jr.: Angew. Chem. 81, 83 (1969); Angew. Chem. Intern. Ed. Engl. 8, 76 (1969).

[64] Vogel, E.: Vortrag auf dem IUPAC-Symposium über Valenzisomerisierungen, Karlsruhe 9.—13.9.1968; zitiert nach [59].

[65] Davis, R. E., Tulinsky, A.: Tetrahedron Letters 1962, 839; J. Am. Chem. Soc. 88, 4583 (1966).

[66] Conrow, K., Howden, M. E. H., Davis, D.: J. Am. Chem. Soc. 85, 1929 (1963).

[67] Anet, F. A. L.: J. Am. Chem. Soc. 86, 458 (1964).

[68] Jensen, F. R., Smith, L. A.: J. Am. Chem. Soc. 86, 956 (1964).

[69] ter Borg, A. P., Kloosterziel, H., van Meurs, M.: Proc. Chem. Soc. (London) 1962, 359; Rec. Trav. Chim. 82, 717, 741 (1963).

[70] Traetteberg, M.: J. Am. Chem. Soc. 86, 4265 (1964).

[71] Butcher, S. S.: J. Chem. Phys. 42, 1833 (1965).

[72] Hoffmann, R. W., Schneider, J.: Tetrahedron Letters 1967, 4347; die ^{1}H-NMR-Messungen wurden von Dr. H. Günther, Köln durchgeführt.

[73] Karplus, M.: J. Chem. Phys. 30, 11 (1959); S. dazu auch die Monographien über NMR-Spektroskopie, z.B. Lit. [5].

[74] Günther, H.: Z. Naturforsch. 20b, 948 (1965). — Günther, H., Wenzl, R.: Z. Naturforsch. 22b, 389 (1967).

[75] Lambert, J. B., Durham, L. J., Lepoutere, P., Roberts, J. D.: J. Am. Chem. Soc. 87, 3896 (1965).

[76] Wei Ch'uan Lin, J. Chin. chem. Soc. 11 (1), 36 (1964); C. A. 61, 9070c (1964).

[77] Günther, H., Görlitz, M.: unveröffentlicht, zitiert in [83].

[78] Gale, D. M., Middleton, W. J., Krespan. C. G.: J. Am. Chem. Soc. 88, 3617 (1966).

[79] Zusammenstellung bei [66].

[80] Abel, E. W., Bennett, M. A., Pratt, L., Wilkinson, G.: J. Chem. Soc. 1958, 4559.

[81] Dunitz, J. P., Pauling, P.: Helv. Chim. Acta 43, 2188 (1960).

[82] Kessler, H., Müller, Eu.: Z. Naturforsch. 22b, 283 (1967).

[83] Günther, H., Görlitz, M., Hinrichs, H. H.: Tetrahedron 24, 5665 (1968).

[83a] Heyd, W. E., Cupas, C. A.: J. Am. Chem. Soc. 91, 1559 (1969).

[84] Knox, L. H., Velarde, E., Cross,A. D.: J. Am. Chem. Soc. 87, 3727 (1965).

[85] Mislow, K., Perlmutter, H. D.: J. Am. Chem. Soc. 84, 3591 (1962).

[86] Mannschreck, A., Rissmann, G., Vögtle, F., Wild, D.: Chem. Ber. 100, 335 (1967).

[87] Dürr, H.: Privatmitteilung — Dürr, H.: Z. Naturforsch. 24b, 1490 (1969).

[88] Tochtermann, W., Schnabel, G., Mannschreck, A.: Z. Naturforsch. 21b, 897 (1966); Liebigs Ann. Chem. 711, 88 (1968).

[89] — Walter, U., Mannschreck, A.: Tetrahedron Letters 1964, 2981 — Tochtermann, W., Schnabel, G., Mannschreck, A.: Liebigs Ann. Chem. 705, 169 (1967).

[89a] Nogradi, M., Ollis, W. D., Sutherland, I. O.: Chem. Commun. 1970, 158.

[90] Tochtermann, W., Stecher, K. H.: Tetrahedron Letters 1967, 3847.

[91] — Horstmann, H. O.: Tetrahedron Letters 1969, 1163.

[92] Childs, R. F., Winstein, S.: J. Am. Chem. Soc. 89, 6348 (1967).

[93] Paquette, L. A.: J. Am. Chem. Soc. 86, 4096 (1964).

[94] Ebnöther, A., Jucker, E., Stoll, A.: Chimia 18, 404 (1964).

[95] Nomenklatur und Bezifferung nach Patterson, A. M., Capell, L. T., Walker, D. F.: The Ring Index, 2nd Edit., Am. Chem. Soc. 1960.

[96] Schönberg, A., Sodtke, U., Praefcke, K.: Tetrahedron Letters *1968*, 3253; Chem. Ber. *102*, 1453 (1969).

[97] Bergmann, E. D., Rabinovitz, M., Agranat, I.: Chem. Commun. *1968*, 334 — Bergmann, E. D.: Bull. Soc. Chim. France *1968*, 1070.

[98] Schönberg, A., Sodtke, U., Praefcke, K.: Tetrahedron Letters *1968*, 3669.

[99] Shimanouchi, H., Ashida, T., Sasada, Y., Kakudo, M., Murata, I., Kitahara, Y.: Bull. Soc. Chem. Japan *39*, 2322 (1966).

[99a] Bertelli, D. J., Gerig, J. T., Herbelin, J. M.: J. Am. Chem. Soc. *90*, 107 (1968).

[100] S. dazu Staab, H. A.: Einführung in die theoretische organische Chemie, 3. Aufl., S. 123 ff. Weinheim/Bergstraße: Verlag Chemie GmbH. 1962.

[101] Kimura, K., Suzuki, S., Kimura, M., Kubo, M.: J. Chem. Phys. *27*, 320 (1957); Bull. Chem. Soc. Japan *31*, 1051 (1958).

[102] Forbes, E. J., Gregory, M. J., Hamor, T. A., Watkin, D. J.: Chem. Commun. *1966*, 114.

[103] Shimanouchi, H., Hata, T., Sasada, Y.: Tetrahedron Letters *1968*, 3573.

[103a] Scherer, K. V., Jr.: J. Am. Chem. Soc. *90*, 7352 (1968) — West, R., Kusuda, K.: J. Am. Chem. Soc. *90*, 7354 (1968).

[104] Bertelli, D. J., Andrews, Th. G., Jr.: J. Am. Chem. Soc. *91*, 5280 (1969) — Bertelli, D. J., Andrews, Th. G., Jr., Crews, P. O.: J. Am. Chem. Soc. *91*, 5286 (1969); dort umfassende Literaturübersicht zum Problem der Nicht-benzoiden aromatischen Verbindungen.

[105] Streitwieser, A., Jr.: Molecular Orbital Theory for Organic Chemists, S. 244. New York: J. Wiley and Sons, Inc. 1961.

[106] Stiles M., Libbey, A. J.: J. Org. Chem. *22*, 1243 (1957).

[107] Naville, G., Strauss, H., Heilbronner, E.: Helv. Chim. Acta *43*, 1221 (1960).

[108] Tochtermann, W., Küppers, H.: Angew. Chem. *77*, 173 (1965); Angew. Chem. Intern. Ed. Engl. *4*, 156 (1965) — Tochtermann, W., Küppers, H., Franke, C.: Chem. Ber. *101*, 3808 (1968).

[109] Siehe dazu [4a], und zwar S. 187.

[110] Tochtermann, W., Franke, C., Schäfer, D.: Chem. Ber. *101*, 3122 (1968).

[111] — Schmidt, G. H.: unveröffentlicht. — Schmidt G. H.: Diplomarbeit, Univ. Heidelberg 1969.

[112] Cristol, S. J., Nachtigall, G. W.: J. Am. Chem. Soc. *90*, 7133 (1968).

[113] Heilbronner, E., Bock, H.: Das HMO-Modell und seine Anwendung, Grundlagen und Handhabung, S. 357. Weinheim/Bergstraße: Verlag Chemie GmbH 1968.

[114] Siehe dazu [4a], und zwar S. 448/449.

[115] Eine ähnliche Annahme macht Schöllkopf, U.: [Angew. Chem. *80*, 603 (1968); Angew. Chem. Intern. Ed. Engl. *7*, 588 (1968)] zur Deutung der Stereochemie der Acetolyse von exo-7-Norcaryl-tosylat.

[116] Williams, J. E., Jr., Sustmann, R., Allen, L. C., Schleyer, P. von R.: J. Am. Chem. Soc. *91*, 1037 (1969).

[117] Tochtermann, W., Horstmann, H. O.: Chem. Ber. im Druck. Dissertation Horstmann, H. O., Univ. Heidelberg 1970.

[118] Streitwieser, A., Jr., Caldwell, R. A., Granger, M. R.: J. Am. Chem. Soc. *86*, 3578 (1964): — Streitwieser, A., Jr., Ziegler, G. R.: J. Am. Chem. Soc. *91*, 5081 (1969).

[119] Lambert, J. B., Oliver, W. L.: J. Am. Chem. Soc. *91*, 7774 (1969).

[120] Paquette, L. A.: J. Am. Chem. Soc. *86*, 4096 (1964).

121) Mannschreck, A., Rissmann, G., Vögtle, F., Wild, D.: Chem. Ber. *100*, 335 (1967).

122) Dittmar, W., Sauer, J., Steigel, A.: Tetrahedron Letters *1969*, 5171.

123) Linscheid, P., Lehn, J.-M.: Bull. Soc. Chim. France *1967*, 992.

124) Nuhn, P., Bley, W.: Pharmazie 22, 532 (1967): — Bley, W., Nuhn, P., Benndorf, G.: Arch. Pharmaz. *301*, 444 (1968).

125) Paul, I. C., Johnson, S. M., Paquette, L. A.: Barrett, J. H., Haluska, R. H.: J. Am. Chem. Soc. *90*, 5023 (1968).

126) Maier, G., Sayrac, T.: Chem. Ber. *101*, 1354 (1968).

127) — Heep, U.: Chem. Ber. *101*, 1371 (1968).

128) Battiste, M. A., Barton, T. J.: Tetrahedron Letters *1967*, 1227.

129) Burchardt, O., Pedersen, C. L., Svanholm, U., Duffield, A. M., Balaban, A. T.: Acta Chem. Scand. *23*, 3125 (1969).

130) Zusammenfassung mit tabellarischer Übersicht. Hall, D. M.: In: Progress in Stereochemistry; Aylett, B. J., Harris, M. M., Hrsg. London: Butterworths 1969; Bd. 4, S. 1.

131) Sutherland, I. O.: unveröffentlicht; zitiert nach 130).

132) Ackermann, K., Chapuis, J., Horning, D. E., Lacasse, G., Muchowski, J. M.: Can. J. Chem. *47*, 4327 (1969).

133) Jackobs, J., Sandaralingam, M.: Chem. Commun. *1970*, 157: — Stevens, J. D.: Chem. Commun. *1969*, 1140.

134) Friebolin, H., Mecke, R., Kabuß, S., Lüttringhaus, A.: Tetrahedron Letters *1964*, 1929.

135) Kurland, R. J., Rubin, M. B., Wise, W. B.: J. Chem. Phys. *40*, 2426 (1964).

136) Oki, M., Hayakawa, N.: Bull. Chem. Soc. Japan *37*, 1865 (1964).

137) — Iwamura, H.: Tetrahedron 24, 2377 (1968).

138) Günther, H.: Tetrahedron Letters *1965*, 4085: — Günther, H., Schubart, R., Vogel, E.: Z. Naturforsch. *22b*, 25 (1967).

139) Tochtermann W., Franke, C.: Angew. Chem. *81*, 32 (1969); Angew. Chem. Intern. Ed. Engl. *8*, 68 (1969).

140) Kabuß, S., Lüttringhaus, A., Friebolin, H., Mecke, R.: Z. Naturforsch. *21b*, 320 (1966).

141) Bredow, K. von: Dissertation Univ. Freiburg/Brsg. 1968. Bredow K. von, Fribolin H., Kabuß, S.: Org. Magnetic Resonance 2, 43 (1970).

142) Moriarty, R. M., Ishibe, N., Kayser, M., Ramey, K. C., Gisler, H. J., Jr.: Tetrahedron Letters *1969*, 4883.

143) Claeson, G., Androes, G., Calvin, M.: J. Am. Chem. Soc. *83*, 4357 (1961): — Scott, D. W., Finke, H. L., Gross, M. E., Guthrie, G. B., Huffman, H. M.: J. Am. Chem. Soc. 72, 2424 (1950): — Scott, D. W., Finke, H. L., McCullough, J. P., Gross, M. E., Pennington, R. E., Waddington, G. I.: J. Am. Chem. Soc. 74, 2478 (1952): — Fehér, F., Schulze-Rettmer, R.: Z. Anorg. Allgem. Chem. *295*, 262 (1958): — Pauling, L.: The Nature of the Chemical Bond, 2nd. Edit., S. 139. Ithaca, N. Y.: Cornell University Press 1960.

144) Nishikawa M., Kamiya, K., Kobayashi, S., Morita, K., Tomie, Y.: Chem. Pharm. Bull. *15* (6), 756 (1967); C. A. *67*, 94826g (1967).

145) Dürr, H.: Z. Naturforsch. *22b*, 786 (1967): — Dürr, H., Tochtermann, W.: unveröffentlicht.

146) Mock, W. L.: J. Am. Chem. Soc. *89*, 1281 (1967).

147) Ammon, H. L., Watts, P. H., Jr., Stewart, J. M., Mock, W. L.: J. Am. Chem. Soc. *90*, 4501 (1968).

148) Williamson, M. P., Mock, W. L., Castellano, S.: J. Magnetic Res., im Druck, zitiert nach l. c. 149).

149) Anet, F. A. L., Bradley, C. H., Brown, M. A., Mock, W. L., McCausland, J. H.: J. Am. Chem. Soc. *91*, 7782 (1969).
150) Tochtermann, W., Franke, C.: Angew. Chem. *79*, 319 (1967); Angew. Chem. Intern. Ed. Engl. *6*, 370 (1967): — Tochtermann, W., Franke, C., Schäfer, D.: Chem. Ber. *101*, 3122 (1968).
151) Michaelis, W., Gauch, R.: Helv. Chim. Acta *52*, 2486 (1969).
152) Fraser, R. R., Schuber, F. J.: Chem. Commun. *1969*, 397, 1474.
153) Schmitt, W.: Vademecum psychopharmacologicum. Karlsruhe: G. Braun, 1968.
154) Bente, D., Hippius, H., Pöldinger, W., Stach, K.: Arzneimittelforsch. *14*, 486 (1964).
155) Bezifferung nach Beilsteins Handbuch der organischen Chemie, Springer Verlag, Berlin, I. E.-W. Bd. 7, S. 212; II. E.-W. Bd. 5, S. 394.
156) *Kessler, H.*: Angew. Chem. *82*, 237 (1970; Angew. Chem. Intern. Ed. Engl. *9*, 219 (1970).
157) *Rauk, A., Allen, L. C., Mislow, K.*: Angew. Chem. *82*, 453 (1970); Angew. Chem. Intern. Ed. Engl. *9*, 400 (1970).
158) Kleier, D. A., Binsch, G., Steigel, A., Sauer, J.: J. Am. Chem. Soc. *92*, 3787 (1970).
159) Sauer, J., Heinrichs, G.: Tetrahedron Letters *1966*, 4979.
160) Heinrichs, G., Krapf, H., Schröder, B., Steigel, A., Troll, T., Sauer, J.: Tetrahedron Letters *1970*, 1617.
161) Combs, C. M., Lobeck, W. G., Jr., Wu, Y.-H.: J. org. Chemistry *35*, 275 (1970).
162) Hoffmann, R. W., Eicken, K. R., Luthardt, H.-J., Dittrich, B.: Chem. Ber. *103*, 1547 (1970).
163) Hoffmann, R.: Tetrahedron Letters *1970*, 2907.
164) Shimanouchi, H., Sasada, Y.: Tetrahedron Letters *1970*, 2421.
165) Prinzbach, H., Vogel, P.: Helv. chim. Acta *52*, 396 (1969).
166) Rahman, R., Safe, S., Taylor, A.: Quart. Rev. (London) *24*, 208 (1970).
167) Russell, G. A., Keske, R. G.: J. Am. Chem. Soc. *92*, 4458, 4470 (1970); dort weitere Lit.

Eingegangen am 3. März 1970

Ergänzungen
(6. August 1970)

Eingehende Diskussionen der Möglichkeiten und Grenzen der Auswertung von temperaturabhängigen NMR-Spektren finden sich auch in den Rotations- und Inversionsprozesse behandelnden Übersichten von Kessler [156] sowie von Mislow et al. [157]. Über eine Möglichkeit zur Umgehung von Fehlern bei der dynamischen NMR-Spektroskopie berichteten Binsch, Sauer et al [158].
Von Sauer [159] und Battiste [128] wurden 1966/67 unterschiedliche Auffassungen hinsichtlich des Vorliegens von Diazanorcaradien- oder Diazacycloheptatrien-Strukturen bei Stickstoff-Heterocyclen vertreten. Neuerdings konnten Sauer und Mitarb. [160] ihre ursprüngliche Zuordnung [159], daß in den Primärprodukten aus s-Tetrazinen und Cyclopropenen Diazanorcaradiene vom Typ *100* vorliegen, mit Hilfe einer Röntgenstrukturanalyse beweisen.

W. Tochtermann

Weitere Literatur

NMR-spektroskopische Untersuchung eines 5-Methylen-5 H-dibenzo |a.d| cyclo-heptens [161].

Konformationsanalyse des 1,7-Dimethoxy-cycloheptatriens [162] MO-Theorie des Cycloheptatrien-Norcaradien-Gleichgewichts [163]

Röntgenstrukturanalyse des Tropolons [164]

Synthese eines stabilen Benzoloxids [165]

Übersicht zur Stereochemie von Polysulfiden [166]

Konformationsanalyse von Semidionen mit siebengliedrigem Ring [167]

ISBN 978-3-540-05101-5   ISBN 978-3-540-36329-3 (eBook)
DOI 10.1007/978-3-540-36329-3

Titel-Nr. 7724

 Springer-Verlag Berlin Heidelberg GmbH

STRUCTURE AND BONDING

Editors: P. Hemmerich, Konstanz; C. K. Jørgensen, Genève;
J. B. Neilands, Berkeley; Sir Ronald S. Nyholm, London;
D. Reinen, Bonn; R. J. P. Williams, Oxford

Structure and Bonding is intended for the publication of papers dealing with problems in all fields of modern inorganic chemistry, chemical physics and biochemistry, where the general subjects are problems of chemical structure and bonding forces.

Vol. 7

With 45 figures
III, 154 pages
(1 contribution
in German) 1970
Soft cover
DM 38,—
US $ 10.50

■ **Prospectus
on request!**

Contents:

The Spectra of Ferric Haems and Haemoproteins. By Dr. D. W. Smith, Chemistry Dept., The University of Sheffield, and Prof. R. J. P. Williams, Inorganic Chemistry Laboratory, Oxford

The Absolute Configuration of Transition Metal Complexes. By Dr. R. D. Gillard and Dr. P. R. Mitchell, Inorganic Chemistry Laboratory, The University, Canterbury, Kent

The Application of Nuclear Quadrupole Resonance. Spectroscopy to the Study of Transition Metal Compounds. By Dr. W. van Bronswyk, "William Ramsey and Ralph Forster Laboratories", University College, Gower Street, London

Kationenverteilung zweiwertiger $3d^n$-Ionen in oxidischen Spinell-, Granat- und anderen Strukturen. Von Dr. D. Reinen, Anorganisch-Chemisches Institut der Universität Bonn